SpringerBriefs in Public Health

For further volumes:
http://www.springer.com/series/10138

Olumurejiwa A. Fatunde • Sujata K. Bhatia

Medical Devices and Biomaterials for the Developing World

Case Studies in Ghana and Nicaragua

 Springer

Olumurejiwa A. Fatunde
School of Engineering & Applied Sciences
Harvard University
Cambridge, MA, USA

Sujata K. Bhatia
School of Engineering & Applied Sciences
Harvard University
Cambridge, MA, USA

ISSN 2192-3698   ISSN 2192-3701 (electronic)
ISBN 978-1-4614-4758-0   ISBN 978-1-4614-4759-7 (eBook)
DOI 10.1007/978-1-4614-4759-7
Springer New York Heidelberg Dordrecht London

Library of Congress Control Number: 2012944562

Printed on acid-free paper

Springer is part of Springer Science+Business Media (www.springer.com)

Preface

The situation surrounding adoption of medical technology in the developing world is a complex and an unsatisfactory one.

Global health has progressed over the last century from a focus on infectious diseases to one on research and aggressive treatment of noncommunicable diseases and improving health care quality. This has precipitated a shift in strategy from implementation of discrete health care interventions, often employing a vertical strategy and targeting a single issue, to transformation of health systems and integration of infectious disease control programs into a strong foundation of primary health care (M Bhatia 2004).

Among the requirements for successful high-level primary health care are health infrastructure on a macro level (health facilities and public amenities that impact access to facilities), competent health personnel, and appropriate technology. Developing countries face unique challenges that affect fulfillment of each of these. Investment in and development of health care infrastructure often lie at the intersection of health and other sectors that fall under a nation's broader development agenda. Substantial research has been conducted regarding the causes and impact of the "brain drain" and other factors that contribute to a consistent shortage of health care personnel, especially in developing countries. Of the three, this book will focus on the development and adoption of appropriate technology and its potential impact on health care delivery.

Medical technology has seen rapid and far-reaching advances in recent decades. However, the development and adoption of most technological innovations have been limited to the industrialized world. Furthermore, "advanced" technology is frequently defined based on a paradigm that considers advancement synonymous with increased digitization and optimization of electronic processes. This book argues that in order for future research to be aligned with the needs of developing and emerging markets, an alternative definition of technology must be embraced. This "new technology" considers naturally derived materials and tools for treatment and includes manipulation of these tools, even in the absence of traditional computerized or electronic technology, under the umbrella of technological advancement.

Once this new definition is established as a reference point, it becomes clear that developing countries have much to offer in terms of contribution to development of new health technologies. Many low-resource communities are at the receiving end of the one-way flow of technologies originating from more industrialized societies, with donation of equipment being the prevailing method of equipment transfer. More often than not, this results in equipment failure or misuse, resulting in failure to expand local capabilities (Kaunomen 2010).

This book will explore the primary reasons for the difficulties that accompany successful transfer of technologies between disparate settings. The book theorizes that the criteria for appropriateness are inherently different in developing societies because of their unique economic and geographic positioning. This volume will begin by delineating the criteria that characterize such settings. The book will then describe the major categories of technology "misfits" and the reason for their failure in developing settings. Additional research into the sources of technological innovation and the factors affecting technology diffusion is needed. Furthermore, the interaction between the private sector, health sector, and regulatory agencies will be a key participant in the organic stimulation of in-country or regional medical technology industries that combine the technological capabilities of multiple sectors and target them towards addressing the technical needs of specific low-resource settings. Formal adoption and standardization of these technologies will then present an additional challenge.

The book will then leave the world of traditional technology and focus on a category that represents an enormous opportunity for developing countries to become active participants in the development of new technologies. As biomedical engineering education and research assume a stronger presence around the world, there is huge potential for the adaptation of biomaterials for clinical use and for their dissemination throughout the developing world and beyond. Biomaterials encompass a range of substances, but one subclassification with high potential is naturally derived and/or synthetically manufactured materials with potential applications in different body systems (Park 1979). Because many of these materials can be found in nature or grown under specific conditions, the agricultural output of developing nations is a natural first instinct for a potential source of naturally occurring biomaterials. The book refers to a special class of naturally derived, scientifically optimized biomaterials as "syntheto-natural" materials with several potential applications.

At this point, the book will undertake a biological analysis of corn- and soy-derived protein mesh materials, particularly with respect to important characteristics such as biocompatibility with human tissues. Similar study of other syntheto-natural materials may result in discovery of other useful biomaterials and inspire developing nations to make agricultural diversification an important priority on their agricultural and larger development agendas (Lin 2011).

The book considers the cases of Ghana and Nicaragua as examples of the broader situation in West Africa and Central America. These two regions are positioned uniquely with regard to both health care and technological capabilities, and both stand to grow significantly in the coming years. While the agricultural sectors of the two nations produce on vastly different scales, both are major producers of corn and other materials that should be investigated further. Together, the scientific and nonscientific considerations discussed here can inform policy decisions that change the face of developing world health care technology.

Cambridge, MA, USA   Olumurejiwa A. Fatunde
                                                                    Sujata K. Bhatia

Acknowledgments

This book was presented as an undergraduate thesis by Olumurejiwa Fatunde to the Harvard School of Engineering and Applied Sciences (SEAS) in May 2012. The following is a statement of gratitude and acknowledgment.

Let me start by thanking God.

The preparation of this document has been a challenging and rewarding journey. My biggest thanks goes to my advisor, Dr. Sujata Bhatia, without whom none of this would have been possible. The completion of the biocompatibility testing discussed herein would have been impossible without the generous assistance of Dr. Manjusri Misra, Dr. Amar Mohanty, and Dr. Nishath Khan at the University of Guelph, Ontario. I'd also like to thank Dr. Anas Chalah, who made doing independent lab work much easier. Thanks also to the many other wonderful people at SEAS and at the Harvard Global Health Institute. I thank my thesis readers, Professor David Cutler and Professor Calestous Juma, for their feedback and guidance, both of which will shape future exploration on this topic. I also thank my professors and others who have inspired me to ponder on problems worth solving.

I owe so much to my family, near and far, for their patience, their tolerance of me never leaving Cambridge because I was "working on my thesis," and their good humor.

Thanks also to my "teammates" William Marks, Dr. Erfan Soliman, and Leslie Rea. And Scott Yim, thanks for the music.

I would like to thank the people who made possible the opportunities that inspired me to write on this topic: Jose Gomez-Marquez and Anna Young of Innovations in International Health (at the Massachusetts Institute of Technology), the Ahoto Partnership for Ghana, Dental Outreach for Africa, and countless others.

I would also like to thank several people without whose advising, mentorship, support, training, encouragement, proofreading, solidarity, etc. this simply would not have happened: Dr. Rehana Patel, Dr. Robin Gottlieb, Dr. Justin Boyd, Dr. Gregg Tucci, Michael Adu, Dr. Tolulope Agunbiade, Kanyinsola Aibana, Henrietta Afari, Chiaka Aribeana, Jessica Ch'ng, Charlotte Chang, Hanna Choi, Veda Eswarappa, Amy Guan, Noam Hassenfeld, Ifedayo Kuye, Dr. Hermioni Lokko,

Naseemah Mohamed, Ngozi Nwaogu, Ekene Obi-Okoye, Olajumoke Odedele, Chinenye Offor, Oludamini Ogunnaike, Oyebola Olabisi, Ifedapo Omiwole, Philip Osafo-Kwaako, Caroline Perry, Stephanie Quaye, Bartholomew Sillah, Linda Ugbah, Clara Yoon, John Yusufu…and the list goes on.

Saira Husain Zaidi, you inspire and amaze me. Your support is a source of constant strength. Thank you.

Timothy Kotin, you make life worth living. Thank you.

Olumurejiwa Adedapo Fatunde
August 2012

Contents

List of Figures

List of Tables

List of Acronyms and Abbreviations

ARC	Austrian Red Cross
BRAC	Development-focused NGO
CGDEV	Center for Global Development
CT	Computed tomography scan
DEVPEM	The development policy evaluation model
DMEM	Dulbecco's modified Eagle's medium
DMF	Dimethylformamide
ECOWAS	Economic Community of West African States
EMEM	Eagle's minimum essential medium
EMP	Medical Service Provider Corporation
EU	European Union
EWH	Engineering World Health
FAO	Food and Agriculture Organization of the United Nations
FAOSTAT	Food and Agriculture Organization Corporate Statistical Database
FDA	Food and Drug Administration
FDB	Food and Drugs Board (Ghana)
GDP	Gross domestic product
GMDN	Global medical device nomenclature
GNP	Gross national product
HAN	Hospital Alemán Nicaragüense
HIV	Human Immunodeficiency Virus
HRC	Hospital Roberto Calderon
INSS	Nicaraguan Social Security Institute
JCLA	Joint Country Learning Assessment
MDR-TB	Multidrug-resistant tuberculosis
MINSA	Ministerio de Salud (Ministry of Health, Nicaragua)
MRI	Magnetic resonance imaging
MTS	Medical technology score
NCD	Noncommunicable diseases
NGO	Non-governmental organization

NHIS   National Health Insurance Scheme (Ghana)
NIC   Nicaragua
OECD   Organization for Economic Co-operation and Development
ORT   Oral rehydration therapy
PAHO   Pan American Health Organization
PHC   Primary health care
PLLA   Poly L-lactic acid
RIGA   Rural income generating activities
SSI   Sustainable Sciences Institute
UNICEF   United Nations Children's Fund
US   United States of America
USD   US Dollars
WHO   World Health Organization

Chapter 1
Introduction

Introduction

Health care in the developing world has improved significantly over the past century and has emerged into a booming sector with plentiful opportunity for growth. The most apparent change has been the shift in focus from infectious disease to noncommunicable diseases (NCDs). Nations face major challenges due to the problems that accompany building and sustaining advanced health systems that provide access to the majority of citizens. Long gone are the days of vertical interventions such as the massive eradication campaigns targeted at smallpox and polio and the less successful malaria campaign of the same nature (Turshen 1989). In the face of chronic infectious diseases such as HIV and those that require intensive, closely regulated treatment, such as tuberculosis, integrated primary health systems have become the vehicle for addressing these issues as well as other increasingly prevalent ailments. However, developing nations have fallen victim to "a high burden of disease despite the availability of basic low-cost preventative and curative measures to avoid premature deaths and disabilities" (Bhatia and Mossialos 2004).

Primary Health Care

The physical manifestation of primary health care (PHC) is a constantly evolving concept. Many developing nations, particularly those impacted by colonial rule, have hospital-based health systems that have traditionally provided inadequate care to the poor. Amidst the health system reform that resulted from the 1978 Alma Ata declaration of the World Health Organization (WHO), a redefinition of health care was intended to precipitate "a shift away from larger hospitals and towards community-based delivery of health services" (Bhatia and Mossialos 2004). Table 1.1 lists areas of redefined focus, with many designed to minimize acute and life-threatening situations.

O.A. Fatunde and S.K. Bhatia, *Medical Devices and Biomaterials for the Developing World: Case Studies in Ghana and Nicaragua*, SpringerBriefs in Public Health, DOI 10.1007/978-1-4614-4759-7_1, © Springer Science+Business Media New York 2012

Table 1.1 Components of
primary health care (PHC)

WHO-defined components of PHC
1. Health education
2. Food supply
3. Proper nutrition
4. Safe water and basic sanitation
5. Maternal and child health care
6. Immunization[a]
7. Prevention of endemic diseases
8. Treatment of common diseases and injuries[a]
9. Provision of essential drugs[a]

[a]Areas where technology can directly affect the
level of care provided
Source: Bhatia and Mossialos (2004)

The success of this transition is dependent on an intentional change in focus and methodology at the policy level, a commitment to the evolution of the supporting foundation, and parallel restructuring of methods of allocating the resources that are necessary to realize a system of this nature. A byproduct of the new focus on PHC is the need for diffusion of health care infrastructure. At the broadest level, this includes construction of health care facilities coupled with improvement of public works (roads, etc.) to allow even the most remote populations to access them. However, even the existing facilities are not performing adequately.

Furthermore, the shift in demographic concentration of the world's poorest societies is surely beginning to have an effect; "urbanization in developing countries is a consequence of rural poverty and deprivation. However, unlike the developed world, where urbanization coincided with industrialization and economic growth in urban settings, poverty in many cities in developing countries remains widespread" (Bhatia and Mossialos 2004).

Health systems are underperforming with regard to standard diagnostic and therapeutic tools, including X-ray machines, basic lab testing and tissue culture facilities, over-the-counter medicines (excluding herbal medications), and treatment of limb injuries surgical tools. For example, patients who seek care at poorly equipped clinics are sometimes asked to provide their own bandages or medications and return to the hospital when they've gathered the supplies necessary for their treatment.

The task for today is to improve the capabilities of existing facilities and to equip them to address the health problems facing their populations. It's important to note that there are different classifications of health facilities in many societies; the vast majority of facilities are not the equivalent of large tertiary health facilities that provide specialized services such as surgery in addition to regular clinical services. Secondary, and especially primary, health care facilities need significant attention in the developing world, particularly because they impact the lives of the greatest percentage of the population.

The lack of established infrastructure means that there is often reduced capacity for inpatient care in local facilities; health care follows more of a consumer product service model in such settings. "The inappropriate introduction of technology does occur in the developing world. However, engineers approach this problem with the assumption that the customers' needs are primary and superior aid results from careful listening to individual doctors (Malkin 2006)." The effect of this assumption is addressed below.

Technology and PHC

Technology has played an important role in medical standardization and has increased the effectiveness of modern health care in industrialized societies. There is now a standard set of equipment that the WHO considers necessary in even the most basic health centers. These pieces of medical equipment fall under the following seven categories: blood transfusion safety, diagnostic imaging, diagnostics and laboratory technology, essential emergency surgical care, injection safety, medical devices and equipment, and transplantation.

Engineering World Health, an organization headed by Duke University professor Robert Malkin, publishes a manual outlining the types of equipment typically found in developing world hospitals (in particular, operating rooms, intensive care units, and emergency rooms) and clinical laboratories (Tables 1.2 and 1.3, Malkin 2006).

A common definition of medical technology, and one that seems consistent with many of today's new devices, is "the use of novel technology to develop highly sophisticated electronic products or medical devices for application in healthcare markets" (Washington Life Science 2012). Increasingly, technologically advanced equipment has allowed for more treatment options that benefit patients and make treatment periods shorter, procedures faster, and patients safer.

For a variety of reasons, technology has not quite reached the same level in less developed regions; as of 2002, 85% of the world's medical devices were produced in the United States, Japan, and the EU (WHO 2003). The WHO lists the following barriers to spread of health care technology to the developing world (WHO 2010d):

- Lack of reliable, clean water, electric power, and adequate public infrastructure
- Gap between standardized health information technology and appropriate implementation
- Low profit margins for device vendors
- Variations in customs and language between countries and even within the same country
- Political instability
- Regulatory constraints and corruption

In the industrialized world, technology is an integral part of health delivery that penetrates almost every level of national health systems. By contrast, most technology that has made its way to developing nations benefits mainly urban areas and the private sector, whose health facilities serve a very small fraction of the total population. In most low-resource settings where there is a mixture of private and public health care, there is a bold line separating the quality of care provided in the two types of facilities.

Table 1.2 Equipment found in the OR, ICU, and ER

Ventilators	Infant warmer
Oxygen concentrators	Phototherapy lights
Fluid pumps	Respiration rate meter or apnea monitor
Electrocardiographs	Electrosurgery machines
Blood pressure machines	Suction machines
Pulse oximeter	Theatre lamps and other lights
Defibrillators	Anesthesia machines
Fetal monitor and fetal Doppler	Bottled gases
Infant incubator	Batteries

Source: Malkin (2006)

Table 1.3 Equipment found in the clinical laboratory

Balances	Autoclaves
Centrifuges and electrical monitors	Laboratory incubators
Microtomes	Water purifiers
Water baths, stir, and hot plates	Clinical laboratory ovens
Microscopes	

Source: Malkin (2006)

Furthermore, care in government facilities is subject to public funding and often cannot support the quantity (number of patients needing care) and quality of care needed to resolve the complexity of medical ailments that patients present with. Very few are keeping pace with the advancements of technology, and many lack even the basic technology that is considered necessary to deliver optimal health care. Of all the medical equipment present in public hospitals, more than 95% is donated from outside sources. A majority of donated equipment is unusable within 5 years of the donation (Malkin 2007b). According to Malkin (2007a), 39% of all imported equipment "never worked due to lack of training, manuals, or accessories." Furthermore, the WHO reports that approximately "70% of medical devices designed for use in the developed world don't work when they reach the developing world" (Malkin 2007a).

This astonishing observation brings to light a set of conflicting observations: the literature cites numerous examples of equipment failing due to limitation of resources within developing countries. On the other hand, Malkin's work and data provided by the WHO reveal that most donated equipment is in fact dysfunctional upon arrival, suggesting that there is a problem inherent in either the donation process itself or in the criteria used to select equipment appropriate for donation. One objective of this book will be to explore the tension between these observations to identify the factors contributing to ineffectiveness of local and imported devices.

In attempting to eliminate this disparity, it becomes important to consider the specific needs and environments of less developed settings. There are medical devices that are simply not needed in the developing world. These include devices that are made to perform highly specialized and unnecessary tasks, and thus can be considered "over-advanced." Many devices in this category are optimized for uses that fall outside of the medical "repertoire" of procedures being performed in the developing

Table 1.4 Classifications of ineffective equipment

	Classification	Example	Comments
A	Outdated equipment that has a newer equivalent	Surgical tools (cauterizing "iron," old wheelchairs)	These tools may not fail immediately, but they often have effects such as lower levels of safety for patients, lower effectiveness of procedures, and non-sterile use. This is less of a problem in the case of wheel chairs, gurneys, and other equipment that may be part of the clinical setup but is not used directly as a treatment method.
B	Malfunctioning "secondhand" equipment that is passed on	Imaging tools (X-ray machines, refurbished microscopes)	This equipment adds to severe health care inefficiencies, and in the case of diagnostics may lower diagnostic capability or even result in misdiagnosis.
C	Equipment with advanced electrical requirements that cannot be supported in the recipient's environment	Temperature-sensitive/ electrically powered machines (incubators, electric lights/cutting tools, blood machines) Advanced imaging: CT, MRI, PET	Equipment in this category is likely to fail and is often among those that are abandoned or cease to function within short periods.
D	Equipment with operational requirements that preclude usage by non-trained workers or that do not provide the training/resources necessary for use	Diagnostic/testing tools (EKGs, paper diagnostics) Advanced imaging: CT, MRI, PET	This may be the largest source of inefficiency; in many cases, this equipment is present and may even be functioning, but another type of barrier prevents it from being beneficial.
E	Electrical/digital equipment with a manual analog that does not sacrifice quality or limit information	Automated blood pressure machine/ sphygmomanometer	

world. In addition, some of the devices that are used in advanced nations are not appropriate in certain settings due to factors such as climate, sterilization needs, etc. Table 1.4 defines five broad classifications of equipment that are inappropriate in developing settings.

As an illustration, consider the example listed under category E of Table 1.4. Below are the average prices for one electrical/manual substitute pair:

Automated blood pressure machine: $101
Sphygmomanometer: $10

Imagine that a blood pressure machine is donated to a small village clinic with the aim of increasing its technological capabilities. Because there is nowhere to plug in its AC adaptor and there is no way to obtain batteries in the immediate area, the machine, while more advanced by some standards, is totally useless in this setting. It may sit unused for months or indefinitely, and the problem of being able to measure and record patients' blood pressure for diagnostic purposes is still unsolved. Did the clinic gain the technological advantage that should have been conferred by automation of this process? For the price of the automated machine that sits on the shelf, the center could have been equipped with ten sphygmomanometers. Ten patients could have their blood pressure checked simultaneously, or additional health facilities in other villages could each be equipped with an additional sphygmomanometer.

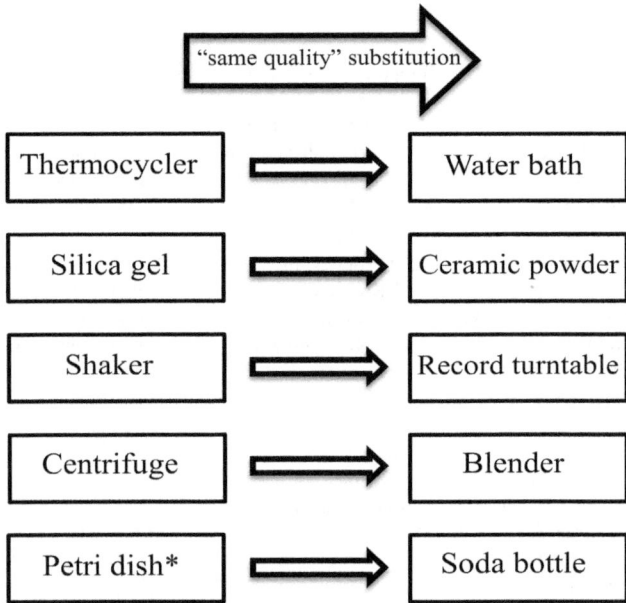

*or other bacteria growth flask

Fig. 1.1 Same-quality equipment substitutions. *Source*: Coloma and Harris (2008)

Other similar substitutes have been identified by the Sustainable Sciences Institute (Coloma and Harris 2008) and include not only relatively simple medical tools such as the sphygmomanometer but also items with household uses that are likely to be commonly found in any setting (see Fig. 1.1).

These ideas provide practical substitutes that could lead to an entirely new class of medical devices. However, we must exercise caution here; Engineering World Health (EWH) aptly warns against the "misconception…that instruments [for the developing world] must be simple" (Malkin 2006). In order to avoid this pitfall, it is important to measure the convenience of these substitute devices along with their efficacy as compared to the original and, above all, their effect (if any) on patient safety. All other things being equal, readaption of the technology involved in these non-medical items (as opposed to recycling of old medical items that have previously been used for their original purposes) presents a much more cost effective and feasible method of designing locally scalable technology.

Because of the overall effects of underdevelopment, the infrastructure needed to support some technologies, such as electricity, temperature control, etc. are not present in some areas. In addition, donation of equipment based on the convenience of the donor, rather than the needs of the recipient (as in the case of categories A and B of Table 1.4), allows technology to be selectively filtered from more to less advanced societies. Unfortunately, the latter are already lagging behind the general advances of technology. The lack of control by recipients over the equipment received poses a dangerous trend that allows external forces to control the rate of local progress. In order to counter this trend, there must be regulation of the donation process and standards that are adhered to by all parties involved (Heydenburg 2008). This is the key to effective transfer and adoption of donated equipment.

It is also imperative that we adjust the current relationship so that recipient facilities and nations determine what they need and what strategies they deem best for optimal use (as in the case of the blood pressure machine). Eventually, as overall technological capabilities continue to expand, developing nations should be able to design equipment based on those needs within their own societies.

The true challenge is that of equipping developing world health facilities with a level of technology that will allow them to meet a certain standard of health provision without blindly imposing the technology of the West, which may or may not be appropriate in low-resource settings. In other words, an independent evolution of culturally, structurally, and economically appropriate technology is needed. This is not intended to dispute the value added by technological innovations that are used in more affluent societies, but rather to emphasize that the research on and development of such items in a given setting is often tailored to the needs and capabilities of that setting, and their success in the home setting does not translate to the success of automatic global dissemination without matching for compatibility.

As things stand, the development and production of medical devices, as well as the formal training of professionals who are able to design, operate, and repair them, has been limited to certain parts of the world. "A major problem with most of the research on technology diffusion is the fact that the developing world is fragmented with a lack of clear power and organization" (Malkin 2007b). Therefore, the market for such products caters to the needs of industrialized nations and parallels the general continued advancement of technology in these societies.

A major difference can be seen between industrialized societies, where traditional technology prevails and is the basis of most innovations, and less industrialized

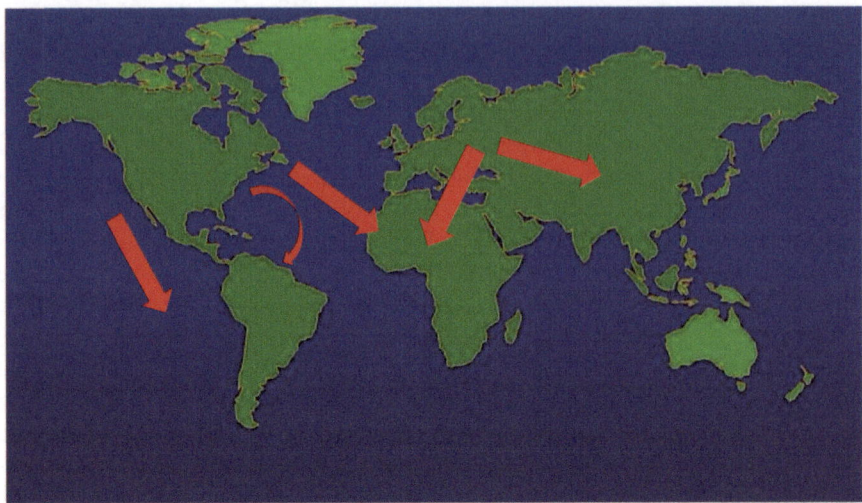

Fig. 1.2 One-way flow of medical technology from industrialized countries to the developing world

societies, where major innovations occur on a smaller and less institutionalized scale. As technical education, particularly in the expanding field of bioengineering, enters previously untouched areas in the developing world, so should the capacity to design technologies that are consistent with local resources. The identification of such resources must be based on the relative strengths of local economies. Until this happens, the current one-way transfer of technology from more to less developed societies (Fig. 1.2), which does little to solve the problem of ineffective care, will persist.

In order to foster autonomy of technological evolution in the societies that need it most, there must be a departure from the traditional definition of technology. For the most part, technological advance has come to be synonymous with computerization, digitization, and building of machines that do things faster and smaller, yet last longer and on a much larger scale. We can articulate a more practical and useful definition of technology by broadening its scope and allowing it to include the use of scientific knowledge to manipulate materials, manmade or otherwise, for the benefit and advancement of a given society.

The current model "displays a characteristic mix of technocratic perspective and equity themes, putting forward health-sector reform in low-income countries simply as an application of reform undertaken in rich countries…health sector reform seems to be driven largely by donor models rather than reform proposals originating within the local health-care system" (Bhatia and Mossialos 2004). In order to move away from said donor model and towards establishing a system in which all countries are on equal footing, we propose the use of the Medical Technology Score (MTS). The MTS has the potential to serve as a language of communication whereby a recipient can communicate its need to the donor, who can then attempt to meet real needs or assist in adapting local resources to meet these needs.

MTS Methodology/Calculation

Determining the "level of technology" in a given nation is a difficult and laborious task, in large part because many of the data needed are ill-defined and their collection is not standardized in developing nations. Comparison to other countries could be meaningless unless the metric is normalized. In order to better understand the needs of specific countries, there is a need for a tool that summarizes the technological standing of a nation according to different parameters. A more detailed discussion of resources that could be used to create such a tool follows in Chaps. 2 and 3 and Appendices B and C. Below is a broad outline of the method that will be used:

MTS: Calculation Method

1. Identify the top causes of local mortality/morbidity
2. Use WHO guidelines for equipment necessary to diagnose and treat these conditions (Hansen et al. 2010)
3. Combine with WHO standards for health care facilities to create an "ideal inventory"
4. Compare national inventory to ideal inventory to obtain the compliance ratio
5. Weight by "essentiality" score of inventory items to create the facility score. Weight average facility scores of different facility categories by density of facilities (WHO 2010c) to create a dimensionless national "score"

The first step should result in the metric being tailored to the situation of the country under study based on its "disease personality." The second reflects the shift in focus towards PHC and a desire to determine whether the level of technology in the nation is suitable for a strong PHC system.

As described above, such a metric requires extensive and accurate national data about classifications of health facilities, regular and comprehensive inventories, and causes of mortality. While information about the latter is available for most nations, many developing world facilities do not currently have the requisite level of detail in their inventories. Ministries of health that wish to prioritize adherence to a national technology standard must take this into account. In Chap. 3, we will use the inventories of health facilities in Nicaragua to create an example of how this metric can be implemented.

The "international standard" to be used in calculating the MTS is the WHO's "Medical Devices by Health Care Facility," a preliminary version of which was recently released (WHO 2011a). The classifications included are given in Table 1.5.

Pending the release of the related document "Medical Devices by Clinical Procedure," a creative approach is needed. For a comprehensive evaluation that takes into account both facility- and disease-specific needs, an individual country might make use of some combination of the following resources:

1. UNICEF Planning Guide for Primary Health Care Centres[1] and First Referral Level Hospitals
2. World Health Organization: Core Medical Equipment

---

[1] "Primary health care center" is here defined as a facility containing the rooms listed in Table 1.6.

Table 1.5 Categories and subcategories of health facilities

Facility type					
	Health post	Health center	District hospital	Provincial hospital	Specialized hospital
Category		Complementary	Complementary	Chemotherapy	Complementary
		Diagnostic	Diagnostic	Complementary	Critical medicine
		Inpatient	Inpatient	Critical medicine	Diagnostic
		Outpatient	Outpatient	Diagnostic	Inpatient
		Treatment	Treatment	Inpatient	Outpatient
				Outpatient	Treatment

Source: WHO (2011a)

Table 1.6 Rooms included in a primary care center

Room/area in a primary care center
Waiting/multipurpose
Toilet (public)
Nurse station
Clean utility
Toilet (staff)
Resuscitation (optional)
Consultation/examination (including reproductive health)
Treatment/dressing/injection
2-Bed room (optional)
Toilet (patient)
Delivery room
Toilet (patient)
Soiled utility/laundry
Primary health laboratory

Source: UNICEF (2005)

3. World Health Organization: Generic Essential Emergency Equipment List
4. Interagency List of Essential Medical Devices for Reproductive Health (WHO et al. 2008)
5. WHO Priority Medical Devices Project: Availability Matrix[2] for Selected Diseases

UNICEF provides clarification on the distinctions between different types of primary health centers:

> A centre that only includes the first few rooms (Treatment, Injection/Dressing, Waiting area) is a nurse station or health post. When the facility also provides consultation/examination and resuscitation services, it is a health centre. Further extending the scope of services results into (sic) the centre becoming a rural hospital (UNICEF 2005).

These classifications, together with the breakdown of national health facilities provided above, can be used to construct a total baseline inventory, weighted by the distribution of facilities in the national health system. Furthermore, we can use the density of health facilities at each level to construct a dimensionless, per capita

---

[2] The availability matrix relies on WHO survey technology to determine the list of preventive, diagnostic, therapeutic, and assistive medical devices needed for the treatment of 15 diseases.

Table 1.7 WHO Availability Matrix diseases

Diseases/conditions	Top ten (Ghana)	Top ten (Nicaragua)
Tuberculosis	X	
HIV/AIDS	X	
Diarrheal diseases	X	
Malaria	X	
Lower respiratory infections*	X	X
Perinatal conditions		
Malignant neoplasms		
Diabetes mellitus		X
Unipolar depressive disorders		
Cataracts		
Hearing loss, adult onset		
Ischemic heart disease	X	X
Cerebrovascular disease	X	X
Chronic obstructive pulmonary disease (COPD)		
Road traffic accidents	X	X

Source: Hansen et al. (2010)
*Pneumonia, lung abscess, and acute bronchitis

metric that can be benchmarked against those of countries that are both much smaller and much larger for comparative analysis.

The WHO Availability Matrix lists the devices that are essential to treatment of the diseases listed in Table 1.7. In compiling a disease-specific inventory for specific nations, only high-impact diseases should be considered. In this text, contributions to national mortality are used as a proxy for relevance of a certain disease. Diseases that are among the top ten causes of mortality[3] in Ghana and Nicaragua are marked with an "X" in Table 1.7.

The Science of Biomaterials

Biomaterials represent a deviation from traditional technology and a new form of technology under the "new definition" established earlier in this chapter. They exist in a field of recent exploration and are unique in the fact that their "technology" comes not from computers or man-made materials, but rather from nature. Malkin suggests that the biggest barriers to design specifically for the developing world include lack of access to spare parts and consumables required for sustained operation (2007a). As resources that can be found locally, biomaterials provide one solution to the challenge of obtaining spare parts. Resources found in developing regions tend to be closer to the "natural" end of the technology spectrum, and developing nations are thus perfectly positioned to take advantage of the possibilities created by development and adoption of technology in this new direction.

---

[3] World Health Rankings (2012).

An important distinction between biomaterials and the biomechanical/electrical tools discussed earlier is that they are a class of medical equipment that requires interaction with human tissues and are not used primarily as an external tool for treatment or clinical diagnosis. For these reasons, the design and selection of bio-materials, particularly those to be used in developing settings, requires additional consideration and research beyond either logistical/cost/efficiency issues or com-patibility with climate and other geography-dependent issues.

Furthermore, advancement of biomaterial use can be integrated into the larger development agendas of developing nations. Agriculture is one of the most promi-nent and fastest-advancing sectors in developing countries. The area where the most synergies can be realized between biomaterial technology and development is likely the study of agriculture at the national level and use of agricultural advances to define the limits and possibilities of biomaterials in these settings. In this book, we will focus on agriculture as a source of biomaterials.

Biological Classification of Selected Biomaterials

There are two major classifications of biomaterials; the term refers to both "such biological materials as tissues and wood" and "selected natural or synthetic implant materials" that "comprise whole or part of a living structure or biomedical device which performs, augments, or replaces a natural function" (Park 1979; San Jose State University 1997). It is this second category which includes numerous sub-stances with potential to affect clinical practice.

In practice, the success of implanting biomaterials is dependent on "the proper-ties and biocompatibility of the implant, the condition of the recipient, and the com-petency of the surgeon who implants and monitors progress of the implant" (Park 1979). The latter two factors will differ widely in each situation, and their optimiza-tion is dependent on a combination of recruiting suitable surgical talent, access to sterile facilities where operations can be performed, and proper evaluation of patient conditions. These tasks are outside the scope of our consideration.

However, it is the task of researchers and biomedical engineers to ensure that the first condition, compatibility between implanted biomaterials and naturally occur-ring biological materials, is met for any material that has the potential to gain wide-spread applicability. In this regard, we see differences arise between natural "reconstituted" materials, such as collagen, and synthetic materials, which can be separated into the four broad categories listed in Table 1.8.

Each of these classifications of synthetic implantable materials has advantages and disadvantages with regard to tolerance of adverse reactions, strength, resilience, and a host of other factors that must all be considered in a study of biocompatibility (Park 1979).

Here we must turn to selection of natural materials with potential for use in implantation. Natural materials have traditionally either been reconstituted prior to clinical use or modified into an alternate structure that allows them to serve new purposes. Until recently, the most commonly used material of this sort was collagen,

Table 1.8 Materials for implantation

Materials	Advantages	Disadvantages	Examples
Polymers Silastic® rubber Teflon® Dacron® Nylon	Resilience, easy to fabricate, low density	Low mechanical strength, time-dependent degradation	Sutures, arteries, vein; maxillofacial—nose, ear, maxilla, mandible, teeth; cement, artificial tendon
Metals 316, 316L S.S. Vitallium® titanium alloys	High-impact tensile strength, high resistance to wear, ductile adsorption of high strain energy	Low biocompatibility, corrosion in physiological environment, mismatch of mechanical properties with soft connective tissues, high density	Orthopedic fixation—screws, pins, plates, wires; intermedullary rods, staples, nails; dental implants
Ceramics Aluminum oxides Calcium aluminates Titanium oxides Carbons	Good biocompatibility, corrosion resistance, inert, high compression resistance	Low tensile impact strength, difficult to fabricate, low mechanical reliability, lack of resilience, high density	Hip prosbook, ceramic teeth, transcutaneous device
Composites Ceramic-coated metal Carbon-coated material	Good biocompatibility, inert, corrosion resistance, high tensile strength	Inconsistent material fabrication	Artificial heart valve (pyrolytic carbon on graphite), knee joint implants (carbon-fiber-reinforced high-density polyethylene)

Reproduced from Park (1979)

which occurs naturally in the body but can also be isolated and treated separately. Collagen substitutes can be found in other species, particularly in bovine animals.

Most early implantable materials, such as poly methyl methacrylate (as used in implanted optical lenses), were discovered accidentally (Ratner and Bryant 2004). Natural materials are unique in that their biocompatibility seems likely, particularly if they are derived from within the same species. The most extreme form of implantation of natural materials is transplant of an entire organ or tissue from another human being. While this practice is quite commonplace in industrialized societies, its use is limited in the developing world because of inadequate facilities, cultural stigma related to use of natural and/or food-associated products in the context of the human body, and other "social, ethical, and immunological problems" (Park 1979). This barrier reveals the existence of a gap and an opportunity for novel discovery of a type of "natural" biomaterial that is compatible with the human body and is readily available in the developing world but is also exempt from the non-scientific barriers that accompany transplantation.

A logical extension of substances that are already part of body tissues are those which regularly come into contact with body tissues, such as those food products consumed by humans that are not found in other animals (i.e., non-meat products such as fruits, vegetables, starches, etc.). These products will be hereafter referred to as agricultural "crops," since they are grown and developed in the agricultural, rather than the laboratory, setting. As we identified earlier, many of the regions under consideration and which are in need of this intermediate "syntheto-natural" product are located in tropical areas of the world, so it is reasonable to limit our discussion to crops that are commonly found in such settings. In this study we will specifically consider corn (maize) and the soybean, two important crops that have tremendous influence on agriculture in the tropical climates of the Western and Eastern Hemispheres, respectively.

Properties of Corn (Maize)

Corn is a grain product that was originally domesticated in Mesoamerica and that spread to Europe and other parts of the world during the trans-Atlantic exploration of the fifteenth and sixteenth centuries. There are several varieties of corn, and the practice of interbreeding and genetic mutation of corn is widespread. Corn is grown in stalks and is a very tall plant in nature. Because of the plant size and the input required to grow the crop, corn that is planted for industrial use is often grown on large plantations. Corn must be planted in early spring, and the plant flowers during the summer, after which it is harvested. Adherence to the growing season is fairly inflexible, as the crop is quite temperature sensitive. Following a harvest of corn, the crop must be stored in a moisture-controlled setting to prevent damage.

A single ear of corn is composed of many kernels, each of which has the following structure (Fig. 1.3):

The most widely studied and used components of corn are gluten and starch, which can be isolated and are often processed and sold as separate products.

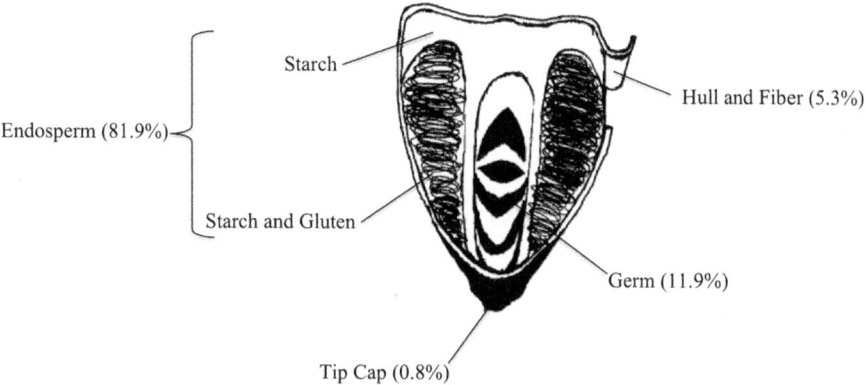

Endosperm (81.9%)

Starch

Starch and Gluten

Hull and Fiber (5.3%)

Germ (11.9%)

Tip Cap (0.8%)

Fig. 1.3 Corn diagram. Based on image from Cereal Food Science

The main difference between the starches of different corn species is the amylose content, which has significant effects on the hardness and gumminess of gels derived from starch. Other properties include swelling power, solubility, turbidity, water binding capacity, gelatinization range (thermal), and other pasting and textural properties (Sandhu and Singh 2007). The potential uses of these materials are strikingly similar to some of the synthetic materials considered for use as biomedical implant materials (Park 1979), and as such it is worth consideration as a possible syntheto-natural substitute. Since corn has traditionally interacted with the human body as a food product rather than a substance in direct contact with the blood-stream and tissues, one factor that must be thoroughly investigated is the biocompatibility of corn starches and corn-derived gels with human tissues and cells.

Properties of the Soybean

The soybean (Fig. 1.4) is a legume native to East Asia. Since its discovery, it has spread across the world, and today it is primarily produced in North and South America. Soybeans are quite varied; they grow in pods of 3–5 on a plant whose height varies by specie.

The major internal components of the bean are oil, protein, and various soluble and insoluble carbohydrates, many of which can be classified as fiber. The soybean is widely considered to be a good source of essential fatty acids due to its high content of essential minerals and vitamins (Plahar 2006). The protein from soybean is relatively heat insensitive, although the bean is sensitive to external moisture and subject to desiccation in the absence of water.

Raw soybeans are toxic to humans, but they are made safe for human consumption via a cooking process whose purpose is to destroy trypsin inhibitors. Products cooked through this process are considered to be valuable alternatives to animal products because they are "complete" proteins that contain all essential amino acids

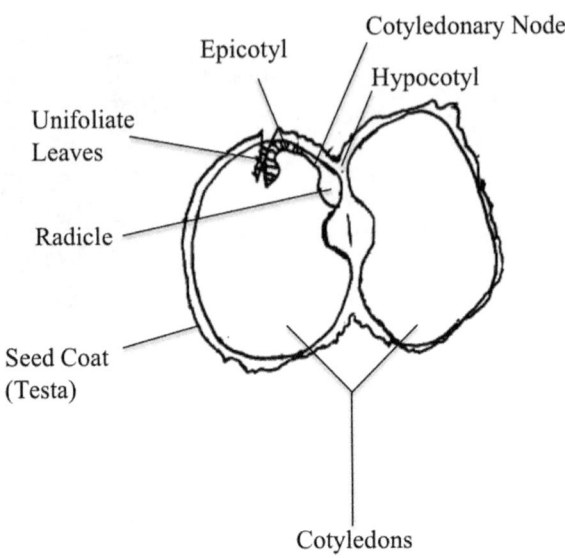

Fig. 1.4 Diagram of a single soybean. Based on image from Texas Tech 2012

(FDA). Soybeans, like maize, are subject to genetic modification. The main commercial uses of the soybean are in the production of oil, soybean meal, flour, and meat substitutes. Because of the abundance of protein in the body and the large proportion that protein occupies as a component of soybean, an examination of soy as a candidate for syntheto-natural implant material would do well to first consider soy protein.

Soy proteins exhibit "elastic, extensible, and sticky properties" (Qi et al. 2011). These proteins are arranged into protein aggregates that are held together by disulfide bonds. Soy proteins can also be fortified with a calcium solution to form a thicker fiber-like material. In fact, the combination of soy with wheat has been tested and suggested as a high-protein alternative to standard wheat bread for developing regions (Mashayekh et al. 2008). These protein aggregates are quite pH-sensitive and are known to denature in high- and low-pH environments. This could have implications for potential areas of application in the human body (Hermansson 1978).

Pending studies of biocompatibility and performance in environments mimicking the human body, corn- and soy-derived biomaterials both present viable options for research into a class of biomaterials that could fill the gap between natural and synthetic implant materials. In this book, we aim to investigate the properties of meshes based on these substances in connection with human cells and to arrive at a preliminary determination of their suitability for such study.

It is important to note that further strides in this direction will have implications not only for the clinical community but also for the agricultural agendas of developing nations. Identification of these materials as targets for future biomaterials may stimulate inquiry into the appropriateness of other prevalent crops and help to underscore the importance of the diversification of agriculture. While economic incentives may motivate individual farmers and entire agricultural systems to focus on their core competency areas, the strategy of diversification could be important in ensuring resilience in the face of climate change, protecting crops such as corn and soy, and making possible the discovery of new crops with the potential for biomaterial implant application (Lin 2011). In the context of health technology, diversification could be a necessary prerequisite for "advancement" under the umbrella of new technology.

Introduction to Focus Countries and Agricultural Connection

In this book, we will limit analysis and discussion to the situation in West Africa and Central America, specifically Ghana and Nicaragua. Each of these nations is a smaller member of its respective region. Ghana and Nicaragua are both part of the Rural Income Generating Activities (RIGA) initiative of the United Nations Food and Agriculture Organization (FAO). According to the World Bank, Ghana is a low-income country, and Nicaragua is a lower-middle income country. In both nations, agriculture has declined as a share of GDP since the 1980s. However, absolute production has increased (at least doubled) as their economies have grown. Comparable countries (in terms of income classification and comparable measures, as determined by RIGA) to Ghana are Bangladesh and Malawi. For Nicaragua, comparable countries are Vietnam and Guatemala (Brooks et al. 2011).

Ghana, with a population of approximately 24 million (about half of whom live in rural areas), is on a strong trajectory of economic growth and is well-positioned to enter a new phase of technological growth and stimulation of industry (Barrientos 2003). Agriculture employs 60% of the total work force and provides income to 70% of the rural population. In addition, 80% of all Ghanaian farmers practice subsistence farming (García 2006; Salisu 2009). Subsistence agriculture is the largest source of income in most communities outside of the major centers of commerce. The crops that subsistence farmers tend to produce for profit, as self-reported by several farmers outside of the Cape Coast region, are maize and corn (Ahoto Partnership for Ghana 2010). Ghana's major agricultural products are cocoa, corn, and sorghum (Department of Agriculture). Further discussion of Ghana in the context of West Africa and the unique nature of its agricultural sector will take place in the next chapter.

By contrast, Nicaragua is much smaller, with a population of almost six million. Technology is more of a regional industry in Central America. This is likely a factor of both the smaller size of these nations and their proximity to the United States. Almost 40% of the Nicaraguan labor force is employed by the agriculture sector; this number has risen since the introduction of commercial farming (Birdsall et al. 2008). Commercial farming commands a larger share of the economy in Nicaragua than in Ghana, and many of Nicaragua's crops are produced for export. The bulk of Nicaragua's agricultural output includes coffee, cotton, bananas, sugarcane, corn, and soya. An in-depth discussion of Nicaragua's focus on PHC and crop diversification efforts will be the subject of Chap. 3.

   Agriculture tends to play a significant role in the economies of developing nations, particularly those in tropical or semitropical regions. Considering the relative strengths of these economies, and in particular the strengths of individuals at the community level, a new "technology" that makes use of materials derived from these crop products seems to be a viable option worth investigation.

Book Outline and Conclusion

The questions driving our exploration are as follows:

1. *What are the major factors limiting the effectiveness of medical technology in developing settings?*
2. *How can the local resources of these settings be harnessed to reverse this trend?*

This chapter has given an introduction to the major problems encountered in developing world health, the role of technology in transforming the health care landscape in these settings, and the need for an alternative definition of technology in order to depart from the current trajectory of technological advance that is appropriate only for the needs of industrialized societies. The next two chapters will focus on the health challenges and agricultural capabilities of the two countries under study. In evaluating the situation of medical technology in each country, we will compare the technical equipment available in specified health facilities with a baseline standard inventory. Chapter 4 will take on a deeper analysis of biomaterials, particularly those derived from corn and soy, and their potential uses in Ghana, Nicaragua, and beyond. Chapter 5 will discuss the process and results of a biocompatibility study of corn- and soy-derived biomaterials. The final chapter will conclude and reflect upon areas for further study in order to fully understand the possibilities discussed herein.

Chapter 2
Case Study of Ghana

Ghana is a prominent West African nation that is widely regarded as having one of the fastest growing economies on the African continent. On a global scale, it is currently emerging into a middle-income nation. It is a key member of the Economic Community of West African States (ECOWAS). Ghana is divided into ten administrative regions (Fig. 2.1). The administration of health services occurs at six levels: community, sub-district, district, regional, tertiary, and national.

Health System Overview

Malaria and measles are leading causes of death in Ghana. Malnutrition compounds the under-five mortality level. While control of infectious disease was a priority for much of the twentieth century, the Ministry of Health and WHO prioritized health system reform and sanitation control in the 1970s and 1980s. Major steps toward this end include the establishment of a national health insurance plan and user fees for health services in 1989 as well as planning for the construction of health centers to make PHC available to 60% of the rural community. Health care financing continues to be a challenge in Ghana; the national budget for health is "channeled disproportionately to existing hospitals and other curative medical care, mainly for the urban minority" (Anyinam 1991). Much of the population, particularly in rural areas, continues to suffer due to lack of access to basic services.

The Ghanaian health sector is pluralistic and has three segments:

1. Home remedy
2. Traditional sector
3. Modern system

Within the modern system, there are several alternatives (Table 2.1).

O.A. Fatunde and S.K. Bhatia, *Medical Devices and Biomaterials for the Developing World:*   19
Case Studies in Ghana and Nicaragua, SpringerBriefs in Public Health,
DOI 10.1007/978-1-4614-4759-7_2, © Springer Science+Business Media New York 2012

Fig. 2.1 Administrative
regions of Ghana

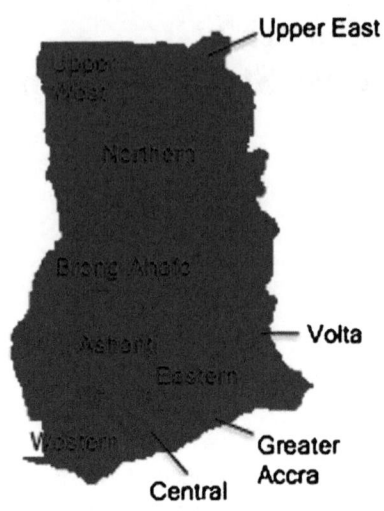

Upper East

Volta

Greater
Accra

Central

Table 2.1 Organization of Ghana's health system

Type of system	Target population	Place of service
Government-operated/financed health system	General population	Hospitals Health centers Clinics Health posts Maternity homes Dressing stations
Quasi-government-operated health care services	Subsets of the population (army, police, certain firms)	
Private medical services offered by individual practitioners	Subsets of the population (based on financial resources)	Hospitals Clinics Pharmacies
Government-supported religious mission health services	Subsets of the population	

Source: Anyinam (1991)

The regional distribution of health services is skewed, as facilities are more heavily concentrated in the south. There is also a disproportionate presence of health facilities (including government-owned centers) in urban areas; according to Anyinam, "over 80% of a sampled 1,000 government-salaried physicians work in the larger towns and cities, which accommodate only 15 percent of the country's population" (1991).

The presence of Western medicine has a long history that is intertwined with the colonial legacy of Britain in Ghana. Much of the health care infrastructure that was established by the English was located in urban areas, particularly in the coastal region, where the majority of European settlements were located (Twumasi 1981). This distribution is still visible in 2012, and it is likely that the health system of colonial times served as the beginnings of today's modern health system. Of the current situation, Asante and Zwi confirm the inequity that is present in different areas of Ghana: "Poverty is greater in the northern sector compared with the southern sector. Similarly, health and healthcare access are unequally distributed, with the poorer regions in the North having the poorest health status" (2009).

The system has also retained its unfortunate practice of best serving the rich: in colonial times, medical services were provided by the Europeans (African doctors were deemed incompetent) and were intended to mainly serve European settlers (Anyinam 1991). Today, top-quality medical services, which are mostly available in private hospitals, are unavailable to the rural population because of logistical constraints and to the poor population (which has a large overlaps with the rural population) because of the prohibitive costs associated with care. The rich, on the other hand, are not only able to afford private care in Ghana, but also often choose to travel to Europe and the Middle East for non-emergency treatment.

The colonial era in Ghana was largely characterized by the lack of a cohesive national health care system. In the 1950s, however, changes began to appear within what would soon become the independent government. Since independence in 1957, the number of health facilities and practitioners has dramatically increased. Paradoxically, the burden of infectious and communicable diseases has also steadily increased throughout this period. Anyinam's explanation for this counterintuitive trend is that "hospital-based, curative services received the greatest attention and too much emphasis was placed on the construction of facilities rather than on the provision of services" (1991). Public health policy was developed to reverse this trend, and in the years following independence the Ghanaian health system began to depart from the colonial model, which did not serve the needs of most people. In the 1970s, the nation made a formal, conscious effort to focus on primary health care. This movement was formalized via the adoption of a primary health care strategy by the Ministry of Health (MoH) in 1978. The program's primary goals were

> to shift the emphasis from the curative to the curative–preventive–promotive approach, from urban to rural populations, from the privileged to the deprived and from a sectoral approach to a system of integrated services as a component of overall social and economic development (Anyinam 1991).

While these were the nominal goals of the initial PHC movement in Ghana, few of the targets for improvement have been met. Since the original announcement of the program, and under the influence of the political regimes that have controlled Ghana since that time, several significant changes have been implemented. The largest of these are the influx of drugs and technology, decentralization of the health care administration, and the introduction of user fees in government hospitals (Anyinam 1991). User fees were initially introduced in an effort to raise funds for public health services, but they quickly began to restrict access to care for the poor (Frempong 2009). In 2003, Ghana passed a National Health Insurance Scheme (NHIS) based on a general tax. As of 2007, 42% of the population was insured under this plan. Private health insurance has not yet infiltrated the Ghanaian market. Asante et al. (2006) demonstrated that in addition to the widespread lack of access experienced by the Ghanaian people, there is a second layer of inequity within regions that subjects the poor to a double burden.

The spectrum of health issues plaguing the Ghanaian populace is somewhat reflective of the health issues across West Africa and in tropical developing nations as a whole.

Technological Capabilities and Realities

A summary of the health care facilities in Ghana, ranging from health posts (the smallest and most rural) to large regional hospitals, which are typically located in urban centers, is shown in Table 2.2.

Health facilities appear to be concentrated at the health center level, with one available to every 10,000 citizens. The major shortages are of skilled health professionals, equipment, and health infrastructure. It has been suggested that the nation should take advantage of its human capital by integrating traditional healers into the modern health system (Anyinam 1991). The use of community health workers and others with limited, yet targeted training is also on the rise. Health care infrastructure is an issue that must be addressed by both the national PHC program and the infrastructure component of the government's economic development agenda. Here we will focus on the changing role and reality of health care technology, particularly as it relates to increasing access for the rural poor, whom have traditionally been neglected in the implementation of health system improvements.

Osei et al. (2005) used the Data Envelopment Analysis (DEA) approach to carry out a pilot study of 17 district hospitals and health centers in Ghana, through which they determined that 47% of the district hospitals and 18% of the health centers studied were technically inefficient. Suggestions for increasing efficiency included converting hospitals to health centers in order to align services provided with capacity and better management of health staff in order to distribute high-level professionals appropriately among health care facilities.

While Ghana does not have a list specifying national standards for medical devices by health facility, the WHO has developed guidelines for standards of medical technology (2011a) as well as inventories of equipment found in health facilities in several nations. Using these as metrics, the goal is to evaluate Ghana's standing against WHO standards and to make possible comparison with economically comparable nations as well as technologically advanced nations that meet WHO standards. We will now develop a general example of the Medical Technology Score (MTS) that encapsulates the major areas of evaluation.

Medical Technology Score

Ghana does not have a unique medical device nomenclature system. The national inventory that is available through the WHO is quite limited, making the calculation of a comprehensive metric difficult. Although Nicaragua will serve as the primary

Table 2.2 Health care facilities in Ghana

Facility	Public sector	Private sector	Total	Density (per 100 k)	% of health care facilities
Health post	285	2	287	1.204	9.55
Health center	1,133	1,231	2,364	9.9172	78.7
District hospital	115	221	336	1.4096	11.19
Provincial hospital	8	0	8	0.0336	0.266
Regional hospital	9	0	9	0.0378	0.299

Reproduced from the WHO Baseline country survey on medical devices (2010c)

Table 2.3 Leading causes of death in Ghana, 2010

Cause	Rank
Diarrheal diseases	1
Stroke	2
Coronary heart disease	3
HIV/AIDS	4
Influenza and pneumonia	5
Tuberculosis	6
Lung disease	7
Malaria	8
Road traffic accidents	9
Kidney disease	10

Source: World Health Rankings

example for MTS calculation in this text, we will discuss a few points that are relevant in the case of Ghana.

Computation of an MTS is more enlightening in the context of comparison to other nations. As such, the metric for different nations must make use of a widely applicable naming system. We will use the Global Medical Device Nomenclature (GMDN). Discussion of this comparison will take place in Chap. 6. In developing this example, the five-step method introduced in Chap. 1 will be of use:

Step 1: Identify the top causes of local mortality/morbidity
According to the most recent available data, the top ten causes of death in Ghana are shown in Table 2.3 (World Health Rankings).

Step 2: Use WHO guidelines for equipment necessary to diagnose and treat these conditions
The bolded entries in Table 2.3 are included in the WHO's Availability Matrix. The devices that are necessary to treat these conditions, along with the devices (in the appropriate quantities) that are necessary to provide a baseline level of primary care in Ghanaian health facilities, would be included in Ghana's baseline inventory for the purpose of MTS calculation.

Determining the necessary scale of technology requires a closer look at the distribution and classification of health facilities in Ghana. The important figures are reproduced in Table 2.4.

Table 2.4 Population served by Ghana's health facilities

Facility	Total	% of health care facilities	Density (per 100 k)	Population served[a]
Health post	287	9.55	1.204	83,000
Health center	2,364	78.7	9.9172	10,000
District hospital	336	11.19	1.4096	71,000
Provincial hospital	8	2.66	0.0336	3,000
Regional hospital	9	3.0	0.0378	3,000

[a]Rounded to the nearest 1,000

Table 2.5 Classification of randomly selected health facilities (example)

Facility	1	2	3	4	5	6	7	8	9	10
Type	HP	HC	HP	PH	HP	RH	DH	DH	HC	HP

HP Health post, *HC* health center, *DH* district hospital, *PH* provincial hospital, *RH* regional hospital

Table 2.6 Actual medical equipment inventory of health facilities 1 through 10 for items A through E (example)

Item	Facility i ($n=10$)									
	1 (HP)	2 (HC)	3 (HP)	4 (PH)	5 (HP)	6 (RH)	7 (DH)	8 (DH)	9 (HC)	10 (HP)
A	x_{1A}	x_{2A}	x_{3A}	x_{4A}	x_{5A}	x_{6A}	x_{7A}	x_{8A}	x_{9A}	x_{10A}
B	x_{1B}	x_{2B}	x_{3B}	x_{4B}	x_{5B}	x_{6B}	x_{7B}	x_{8B}	x_{9B}	x_{10B}
C	x_{1C}	x_{2C}	x_{3C}	x_{4C}	x_{5C}	x_{6C}	x_{7C}	x_{8C}	x_{9C}	x_{10C}
D	x_{1D}	x_{2D}	x_{3D}	x_{4D}	x_{5D}	x_{6D}	x_{7D}	x_{8D}	x_{9D}	x_{10D}
E	x_{1E}	x_{2E}	x_{3E}	x_{4E}	x_{5E}	x_{6E}	x_{7E}	x_{8E}	x_{9E}	x_{10E}

The "first referral level hospital" (a UNICEF classification) is described by UNICEF (2005) as a hospital at a district level, or a facility that serves 50,000–500,000 people. Based on the density of health facilities in Ghana, we would consider district, provincial, and regional hospitals as part of this category.

Step 3: Combine with WHO standards for health care facilities to create an "ideal inventory"

As an example, consider ten randomly selected health facilities in Ghana. Each of these can be assigned to one of the five WHO facility categories based on its size, the services offered, and the surrounding population characteristics, such as density (Table 2.5).

Each of the ten hospitals has a certain number of items A through E, which should be recorded in an inventory. A combined inventory of the ten hospitals might resemble Table 2.6.

Table 2.7 Ideal equipment inventories of the five health facility categories for items A through E (example)

| Item | Facility type | | | | |
	HP	HC	DH	PH	RH
A	$y_{\text{HP-A}}$	$y_{\text{HC-A}}$	$y_{\text{DH-A}}$	$y_{\text{PH-A}}$	$y_{\text{RH-A}}$
B	$y_{\text{HP-B}}$	$y_{\text{HC-B}}$	$y_{\text{DH-B}}$	$y_{\text{PH-B}}$	$y_{\text{RH-B}}$
C	$y_{\text{HP-C}}$	$y_{\text{HC-C}}$	$y_{\text{DH-C}}$	$y_{\text{PH-C}}$	$y_{\text{RH-C}}$
D	$y_{\text{HP-D}}$	$y_{\text{HC-D}}$	$y_{\text{DH-D}}$	$y_{\text{PH-D}}$	$y_{\text{RH-D}}$
E	$y_{\text{HP-E}}$	$y_{\text{HC-E}}$	$y_{\text{DH-E}}$	$y_{\text{PH-E}}$	$y_{\text{RH-E}}$

Table 2.8 Ideal equipment inventory of health facilities 1 through 10 for items A through E (example)

| Item | Facility i ($n=10$) | | | | | | | | | |
	1 (HP)	2 (HC)	3 (HP)	4 (PH)	5 (HP)	6 (RH)	7 (DH)	8 (DH)	9 (HC)	10 (HP)
A	y_{1A}	y_{2A}	y_{3A}	y_{4A}	y_{5A}	y_{6A}	y_{7A}	y_{8A}	y_{9A}	y_{10A}
B	y_{1B}	y_{2B}	y_{3B}	y_{4B}	y_{5B}	y_{6B}	y_{7B}	y_{8B}	y_{9B}	y_{10B}
C	y_{1C}	y_{2C}	y_{3C}	y_{4C}	y_{5C}	y_{6C}	y_{7C}	y_{8C}	y_{9C}	y_{10C}
D	y_{1D}	y_{2D}	y_{3D}	y_{4D}	y_{5D}	y_{6D}	y_{7D}	y_{8D}	y_{9D}	y_{10D}
E	y_{1E}	y_{2E}	y_{3E}	y_{4E}	y_{5E}	y_{6E}	y_{7E}	y_{8E}	y_{9E}	y_{10E}

A standard WHO inventory that delineates which items should be present in which quantities for a given type of facility is illustrated in Table 2.7.

x_{iA} is the actual quantity of item A in hospital $i=1\ldots n$ and y_{iA} is the ideal quantity (determined using UNICEF guidelines) of item A for the given hospital type. The classifications in Table 2.5 can be used to map the ideal quantities $y_{(\text{facility})A}$ to each of the ten facilities, resulting in an ideal inventory for each facility (see example below).

Example:

Facility 1 is a health post. Therefore, the ideal inventory for facility 1 should be equal to the ideal inventory of a health post:

Item	Facility 1 (HP)
A	$y_{1A}=y_{\text{HP-A}}$
B	$y_{1B}=y_{\text{HP-B}}$
C	$y_{1C}=y_{\text{HP-C}}$
D	$y_{1D}=y_{\text{HP-D}}$
E	$y_{1E}=y_{\text{HP-E}}$

Application of this method to each facility in turn results in the facility-specific ideal inventory (Table 2.8).

Step 4: Compare national inventory to ideal inventory to obtain the compliance ratio

Step 5: Weight by "essentiality" score of inventory items to create the facility score. Weight average facility scores of different facility categories by density of facilities (WHO 2010c) to create a dimensionless national "score"

We can then define the score for item A as

$$score_A = \frac{x_A}{y_A} \tag{2.1}$$

$$score_{hosp} = \frac{\sum\limits_{i=A,B,C,D,E} (score_i)(e_i)}{n} \tag{2.2}$$

where $score_{hosp}$ is the score for the entire hospital or unit being considered, n is the total number of items in the hospital ($n=5$ in this case), and e_i is the measure of "essentiality" for a given item.[1]

For each category of health facility, we can assign an aggregate score that is the simple average of each hospital's score:

$$score_c = \frac{\sum\limits_{i=1}^{N_c} score_{hosp,i}}{N_c}, \tag{2.3}$$

where N_c is the number of facilities in the category (provided by the WHO Baseline country survey on medical devices) for a given nation.

Formulation of a single score for a nation based on its five category scores will involve qualitative and quantitative assessments that take into account several factors, such as which facilities provide the majority of treatment, population distribution, and relative facility distribution, etc. Thus the ability of the MTS to approximate reality is very much limited by the availability of data for these related factors. In spite of this limitation, an attempt at a more complete example is provided in Chap. 3 using inventories from Nicaraguan hospitals.

Agriculture

In addition to health care, agriculture is another sector that has been identified as a priority in the context of Ghana's continued growth and advancement. As 70% of the rural population (which is itself a majority of the Ghanaian population) is involved in agricultural activities as a form of livelihood, there is real opportunity to both generate revenue and increase the standard of living by making the sector more effective. National trade policy has recently shifted the balance of exports and imports in the agricultural sector (Abdulai and Egger 1992).

---

[1] This measure is based on the Fenningkoh and Smith model (Appendix B), which uses a point system to evaluate items based on function, risk, and required maintenance (*source*: WHO Medical device technical series).

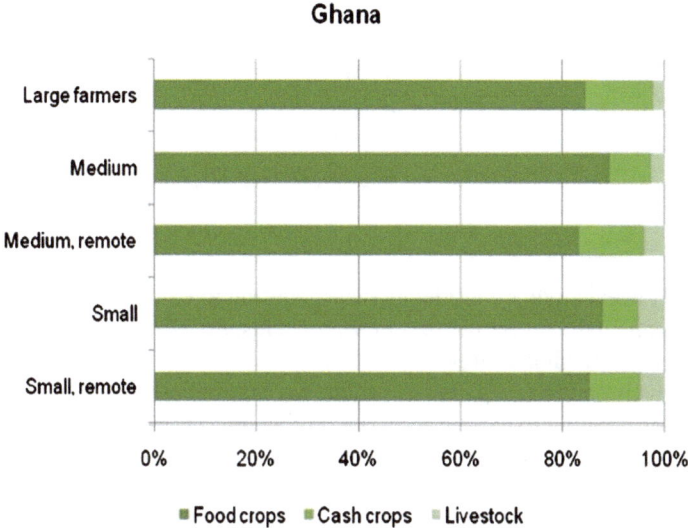

Fig. 2.2 Farm household production mix—Ghana (Brooks et al. 2011)

At the household level, crop production is the major source of agricultural income (Fig. 2.2).

Agricultural aid in Ghana has increased since 2000. The largest area of aid is agricultural production. From 1964 to 1983, growth in the production of cereals was mostly due to expansion of agricultural land area. However, from 1984 to 2004, the growth was primarily driven by improvements in the yield of existing production (Dewbre and de Battisti 2008). This shift is indicative of the increasing effectiveness of farming methods. However, continued expansion of cultivation area must continue to be a priority in order to achieve maximal scale and capitalize on the comparative advantage conferred by the climate and abundance of resources in Ghana. As of 2006, it "was reported … that only 5 per cent of irrigable land in Ghana is actually irrigated, leaving production largely dependent on the vagaries of the weather" (African Development Bank and OECD 2007).

Duku et al. have explored the potential use of biomass resources as biofuels in Ghana. Biomass is defined as "all organic matter that is derived from plants as well as animals. Biomass resources include wood and wood wastes, agricultural crops and their waste by-products, municipal solid waste, animal wastes, wastes from food processing, aquatic plants and algae" (2011).

Given the prevalence of subsistence agriculture, significant attention has been given to the collectivization and commercialization of agriculture. In particular, there is support behind a shift from growing subsistence crops, which include maize and corn, to production of Ghana's major agricultural products (cocoa, corn, and sorghum) (Department of Agriculture). These changing dynamics are reflective of a fundamental conflict between two ideas which are based on divergent priorities. Commercialization with the goal of increasing national output lends support to the idea of developing competitive advantage in the production of few, high-yield

Table 2.9 Components of
soybean processing

Large-scale processing of soybean	
Animal feed production and oils extraction	55%
Soy flour and high protein foods	20%
High protein foods only	15%
Soymilk and soy flour	5%
Soymilk and soy curd	5%

Source: Plahar (2006)

products. However, other considerations are starting to emerge, including the risks and environmental dangers that accompany exclusivity, as well as the potential of other uses for agricultural output.

Our focus is on this last point—given the agricultural output of a nation, there is a possibility, pending extensive research on the subject, that some portion of the nation's output of corn-like products could be redirected into the production of bio-materials. As maize is mostly produced at the subsistence level, confirmation of its potential as a biomaterial would provide additional support for the scaling up of corn production and diversification of the national agricultural production palette.

The six major types of agro-ecological zones in Ghana are the rain forest, deciduous forest, transition, Guinea savannah, Sudan savannah, and Coastal savannah. Maize is a major food crop in the transition zone and the Guinea and Coastal savannahs, which together make up 93% of the total land area (Duku et al. 2011). Maize and the other major crops are largely produced via manual cultivation.

Maize is the fourth largest crop in production, as measured by quantity and using local prices (FAOSTAT). In addition to being a crop that is produced and exported on a national level, maize is one of the crops most frequently produced by subsistence farmers (Ahoto Partnership for Ghana 2010). Logistically, national production of maize is the most likely target for allocation of maize for medical purposes.

Soy was introduced to Ghana in 1910, but struggled commercially throughout the twentieth century due to the lack of a strong industrial base for soy processing and a very limited market for the crop (Plahar 2006). In recent years, demand has increased due to changes in Ghanaian habits. The major form of consumption is soy sauce. While production increased in the 1970s in response to a soybean campaign, most of the soy production in Ghana happens in the northern regions (Dzogbefia et al. 2007).

In the 1980s and 1990s, the Ministry of Agriculture formed a National Committee on Soybean Production and Utilization which included input from several government ministries, industry experts, and food and agriculture professionals. There was a concerted effort to increase production of soybeans in Ghana, complete with plans for monitoring changes in production and post-production processing. Additional emphasis was placed on promoting industrial and household utilization of soybean products. In Northern Ghana, the major destinations of soy products are food for animals and humans. Soy products in southern Ghana are mostly derived from crops grown in the North, and the products are passed from retailers to late-stage processors, who then transmit products to end consumers (Dzogbefia et al. 2007).

Existing soybean processing techniques are largely inefficient and, in some cases, inappropriate. Training programs have been developed for women and schoolteachers

Table 2.10 Crop production in Ghana

Country	Item	2009 Production (tons)
Ghana	Cocoa beans	662,400
	Coffee, green	1,500
	Maize	*1,619,590*
	Maize, green	
	Millet	245,550
	Oil palm fruit	2,103,600
	Rice, paddy	391,440
	Sorghum	350,550
	Soybeans	–
	Sugar cane	145,000
	Coarse grain, total + (total)	2,215,729
	Fiber crops primary + (total)	7,883

Source: FAOSTAT

in order to spread knowledge about the cultivation and use of soy. Large-scale processing of soy in Ghana is broken down into several components (Table 2.9).

Thus far, the focus of the soy campaign has been on the promotion of household utilization and development of new recipes and products, including those based on combinations of soy and other abundant crops (e.g., maize-soy blends). However, challenges such as long cooking times and unfamiliarity with the taste of the bean have proven to be obstacles to its spread in the Ghanaian market. Furthermore, anti-nutritional factors, potential negative side effects, and the need to follow a stringent cooking regimen in order to remove toxic trypsin inhibitors are additional barriers to usage. Despite production numbers that are in line with other major export crops, soybean consumption is merely 0.03% of soy produced in the country (Dzogbefia et al. 2007).

Such low levels of utilization make it difficult to sustain high levels of production, feeding into a vicious cycle that is consistent with the low levels of soy in use today. Despite the difficulties associated with increasing domestic soybean usage, Ghanaian authorities recognize the potential benefit the soybean would confer as a cash crop as well as its usefulness as a raw material in the oil, livestock, and poultry industries (Plahar 2006). The incentive to continue encouraging consumption is therefore quite high.

This gap between household usage and production exposes a perfect opportunity for alternative usage of soybeans in Ghana. The introduction of soy-derived biomaterials is one such possible alternative; it provides both a steady avenue for utilization that is independent of societal consumption norms (changes to which require an extended time scale) and motivational for additional production efforts.

Data from the Food and Agriculture Organization Corporate Statistical Database (FAOSTAT) provides an overview of the relative levels of production for various crops in Ghana (Table 2.10).

It is evident that Ghana's major items of production are maize and oil palm fruit. Furthermore, soy is not produced on a commercial scale in Ghana. However, the landscape for the soybean changes if we consider Ghana's immediate neighbors (Burkina Faso, Cote D'Ivoire, and Togo) (Table 2.11).

Table 2.11 Production of maize and soybean in Ghana and neighboring countries (2009)

Item	Country	2009 Production (tons)
Maize	Burkina Faso	894,558
	Côte d'Ivoire	680,000
	Ghana	1,619,590
	Togo	651,738
	Total	3,845,886
Soybeans	Burkina Faso	15,686
	Côte d'Ivoire	679
	Ghana	
	Togo	
	Total	16,365

Source: FAOSTAT

Table 2.12 Production of maize and soybean in West Africa (2009)

Maize		Soybean	
Country	2009 Production (tons)	Country	2009 Production (tons)
Benin	1,205,200	Benin	57,308
Burkina Faso	894,558	Burkina Faso	15,686
Cape Verde	7,380	Cape Verde	
Côte d'Ivoire	680,000	Côte d'Ivoire	679
Gambia	54,625	Gambia	
Ghana	1,619,590	Ghana	
Guinea	565,660	Guinea	
Guinea-Bissau	14,000	Guinea-Bissau	
Liberia		Liberia	3,000
Mali	1,476,990	Mali	2,625
Mauritania	12,497	Mauritania	
Niger	1,389	Niger	
Nigeria	7,338,840	Nigeria	573,863
Senegal	328,644	Senegal	
Sierra Leone	29,641	Sierra Leone	
Togo	651,738	Togo	
Total	14,880,752	Total	653,151

Source: FAOSTAT

Consideration of the entire West African region sheds further insight into the availability of soy through intra-regional trade (Table 2.12).

This data provides a measure of confidence that a large pool of potential corn- and soy-derived biomaterials is available to the Ghanaian health sector based solely on agricultural output.

Ghana is a member of the ECOWAS, which confers benefits in the form of lower tariffs on intra-regional trade. The current structure of tariffs in Ghana conforms to the breakdown in Table 2.13.

Table 2.13 Ghana tariff schedule

Item category	Rate (%)
Primary products, capital goods, and some basic consumer goods	0–5
Raw materials, intermediate inputs, and consumer goods	10
Final consumer goods	20

Source: Dewbre and de Battisti (2008)

The effect of these tariffs is to benefit domestic producers of imported goods in the listed categories. Despite Ghana's ECOWAS membership, "trade with other member countries....tends to be small compared to trade with non-members, principally OECD member countries" (Dewbre and de Battisti 2008). Use of crops for nonagricultural purposes may provide additional incentive to consider shifting the balance of trade within and outside the region.

Following an investigation of human biocompatibility, an important area of necessary research is the translation of these materials from agriculture to the health sector and the quantification of the need for biomaterials, as a substantial diversion of resources will certainly impact future agricultural policies.

Discussion and Conclusion

This chapter has outlined the major health problems that plague the West African nation of Ghana, the situation of health technology in Ghana, and the aspects of the agricultural sector relevant to biomaterial production.

Ghana is faced with the health challenges common to many sub-Saharan African nations; while infectious diseases still factor heavily into the burden of disease, an emerging focus on primary health care and treatment of chronic disease has marked the last few decades. Inequity in access to health services is still prevalent both between and within regions of the country. However, Ghana is uniquely positioned for vast economic growth. With proper planning and coordination among different sectors, this can translate into new strides in the health sector and progress towards the goal of access for all Ghanaians.

Development and publication of detailed equipment inventories will be necessary in order to calculate a MTS for Ghana. However, an initial look at health statistics on Ghana indicates that the majority of Ghanaian health facilities are mid-sized health centers, of which there is one for every 10,000 Ghanaian citizens. Analysis of the realities of health technology in Ghanaian health facilities will allow for comparison to nations with similar profiles and will inform next steps for maintaining a level of technology that reflects Ghana's needs and its commitment to expanding PHC.

Ghana's agricultural sector is characterized by a large number of subsistence farming units, which supplement a more centralized national output. Among the biggest producers of agricultural income are cocoa (the nation's top exported good) and maize, which is a net imported good despite its high levels of production.

While the soybean is not a major agricultural product in Ghana, it is produced in the northern regions of the nation as well as in surrounding countries. The use of soy in biomaterials provides an additional outlet for utilization of current soy production and an incentive for further scale-up of production, particularly in the coastal regions.

Ghana's regulatory body, the Food and Drugs Board, exercises control over the export and import of both food products and medical devices (FDB). According to the updated fee schedule for 2010, local medical devices are subject to a fee of 600 Ghana Cedis (equivalent to USD 341.66), while foreign medical devices are imported for a fee of USD 1,500. The tariff structure favors the development of local devices that can be disseminated via regional trade.

Adoption of a biomaterial sector in Ghana based on corn- and soy-derived materials is feasible based on existing conditions. However, for optimization of such an industry, more attention will need to be given to the production of the soybean. Further inquiry into the logistical and regulatory challenges surrounding this issue will inform the plan of action for the collaboration between the agricultural and health sectors in years to come.

Chapter 3
Case Study of Nicaragua

Nicaragua is a nation situated southwest of the island of Hispaniola and nestled in between the Pacific Ocean and the Caribbean Sea. With a population of just under six million, Nicaragua is the second least-populated nation in Central America. Nicaragua is a member state of the Pan American Health Organization (PAHO), the Regional Office of the World Health Organization (WHO) for the Americas.

The geographic proximity of Nicaragua to two major water bodies results in a natural division of the nation into Pacific and Atlantic regions as well as a Central region, which serves as a buffer between the two. The land area and population are distributed among the three regions as shown in Table 3.1.

Although it represents less than one-fifth of the total land area, the Pacific region contains the majority of the population. The nation is formally divided into 15 administrative regions and 2 autonomous regions (see Fig. 3.1).

The nation was severely affected by political strife in the late twentieth century. As a result of the Somoza insurrection (ending in 1979), significant damage was sustained on the Pacific (west) coast of the nation. Reconstruction and subsequent development of the health system have resulted in two very different models of health care being practiced on the two coasts since 1979, when Sandinista came to power (Donahue 1991). Of all the nations in the Mesoamerican region, Nicaragua has the second-lowest growth rate in terms of year-over-year GDP growth during the period from 1990 to 2003 (OECD 2006).

Health System Overview

The health conditions in Nicaragua are largely tied to vector-borne and communicable diseases that thrive in the tropical climate of Central America. Dengue, leishmaniasis, and Chagas' disease are all major concerns, while the prevalence of malaria and tuberculosis has steadily declined over the past decade (PAHO 2007). Due to improvements in Nicaraguan immunization strategies, vaccine-preventable diseases have been largely controlled, with some such as diphtheria having been

O.A. Fatunde and S.K. Bhatia, *Medical Devices and Biomaterials for the Developing World:*   33
Case Studies in Ghana and Nicaragua, SpringerBriefs in Public Health,
DOI 10.1007/978-1-4614-4759-7_3, © Springer Science+Business Media New York 2012

Table 3.1 Division of regions in Nicaragua

Region	% of total land area	% of total population
Pacific	15.2	54
Central	28.4	32
Atlantic	56.4	14

Source: PAHO (2007)

Fig. 3.1 Administrative
regions of Nicaragua

totally eliminated from the population (PAHO 2007). The major challenges currently facing the nation are acute respiratory infections, noncommunicable and cardiovascular diseases, and malnutrition. This parallels a trend in developing societies
worldwide.

Since the 1980s, the theme of decentralization, a strong immunization program
including National and Regional Health Days focused on 100% coverage, and
movement from an institutional/professional model to a popular model have characterized the Nicaraguan health system. Notable in this shift has been the training
and usage of Community Health Workers, commonly known as "Brigadistas"
throughout the nation, but especially in the less densely populated Atlantic region.
The Brigadista model relies heavily on inter-training of workers for self-perpetuation,
and it has been a successful solution to the general lack of human resources, especially
in rural areas (Donahue 1991).

The disparities in care and different models employed on the Pacific and Atlantic
coasts are significant, and they are still manifested in the health situation of today.
The areas with highest child malnutrition are RAAN, Jinotega, Madriz, and
Matagalpa, which are all located in the north and in the Atlantic or Central Regions
(PAHO 2008).

The Pacific Coast is characterized by an institutional primary health care plan as
well as vertical integration of services and popular participation. The Atlantic
Coast, on the other hand, generally provides fewer services. As of the early 1990s,

Fig. 3.2 Components of the
health system

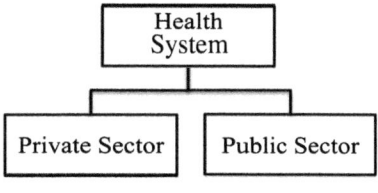

Fig. 3.3 Components of the
private sector

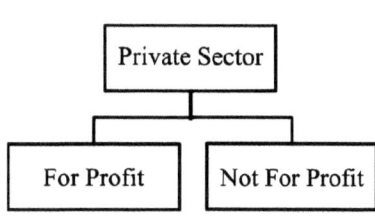

many services were provided by Protestant churches in the area. The Brigadista network was one of the main sources of service on the Atlantic coast and remained in place even after Sandinista's rise to power following the revolution. The remnants of these historical causes of health inequity will be important to consider in the context of modern challenges and possible opportunities to collaborate with the agricultural sector.

Populations of Central American nations tend to be loosely negatively correlated with average income and literacy and positively correlated with malnutrition and infant mortality. Nicaragua is second to Costa Rica in lowest population, lowest malnutrition, and highest literacy. However, it has the lowest GNP per capita in the region (Barrett 1996). This coexistence of poverty and relatively strong health indicators give the nation a unique position among its neighbors.

Nicaragua also has the second highest number of physicians per capita in Central America. In the late 1970s, 23 health agencies in Nicaragua were integrated into a single national organization following the change of government. As of the mid-1980s, Nicaragua had the highest health expenditures as a percentage of GNP in all of Central America. There was also quite an increase in immunization levels, partially due to the consolidation of health agencies and institutionalization of popular health days (Barrett 1996). As of 2008, health expenditures were 9.4% of the GDP, and health expenditures per capita were USD 105 (WHO 2011c).

The organization of the Nicaraguan health system is shown in Fig. 3.2.

The private and public sectors are further divided into subunits and interlinked departments as shown in Figs. 3.3 and 3.4.

Insurance is largely provided by the Nicaraguan Social Security Institute (INSS), which contracts with local Medical Service Provider Corporations (EMPs) to provide services to people. Responsibility for health care provision is split between the Ministry of Health, INSS, private services, and military/civil service providers (PAHO 2007).

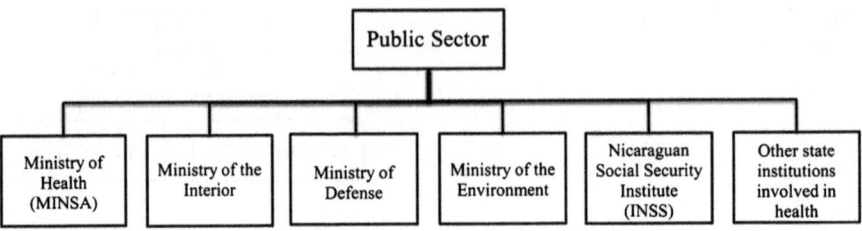

Fig. 3.4 Components of the public sector

Table 3.2 Distribution of health care services in Nicaragua by facility

Type of facility	Types of care offered
Outpatient care services	Preventive programs
	Outpatient visits
	Emergency visits
Hospitals	Emergency care
	Inpatient care
Secondary care	Departmental hospitals
	National referral hospitals
Tertiary care	National centers for cardiology, radiotherapy, ophthalmology, dermatology, psychiatry, and laboratory services

Source: PAHO (2007)

Table 3.3 Health care facilities in Nicaragua

Facility	Public sector	Private sector	Total	Density (per 100 k)	% of health care facilities
Health post	890	N/A	890	15.4977	80.47
Health center	158	N/A	158	2.7513	14.29
District hospital	26	N/A	26	0.4527	2.35
Provincial hospital	26	N/A	26	0.4527	2.35
Regional hospital	6	N/A	6	0.1045	0.54

Reproduced from the WHO Baseline country survey on medical devices

In 2004, the Nicaraguan government put into place a National Health Plan that will be in effect until 2015 (PAHO 2007). The main features of the plan are commitment to increased capacity and a stronger primary health care system.

The range of health care facilities operated in Nicaragua includes hospitals, general and maternity clinics, and NGO-operated clinics. Services are distributed according to Table 3.2.

A summary of the health care facilities in Nicaragua is given in Table 3.3.

Over 80% of all health care facilities in Nicaragua are health posts, and 95% of facilities are health centers or health posts. This suggests a slightly higher level of local penetration than in Ghana, where these two facility types comprise only 88% of health facilities. Interestingly, the balance between health posts and health centers appears to be inverted for the two nations, with Nicaragua having an overwhelming majority of health posts and Ghana having more health centers. Both are rural establishments, but health posts are often remote and in the midst of specific towns or communities, while health centers serve multiple towns and might be considered the smallest facilities affiliated with a conglomerate of communities or towns. This difference in balance may be due to both the much larger population in Ghana and the presence of regional governments in each of Ghana's ten administrative regions, suggesting an "upward shift" towards urbanity. The density in Table 3.3 suggests that a health post exists for every 6,453 people, but the disparities in access on the two coasts of Nicaragua belie that statistic.

In a fashion mirroring the disparities in Ghanaian health provisions and outcomes between the coastal and inland regions, there are significant structural differences between health care delivery on Nicaragua's Pacific (west) coast and its eastern Caribbean-Atlantic coast (Donahue 1991). Unlike Ghana, these differences can be attributed not to an external colonial legacy, but to change in political leadership and internal strife.

Technological Capabilities and Realities

PAHO and the Ministry of Health have identified a number of shortcomings concerning the technical capacity of the Nicaraguan health system. According to PAHO's 2007 report,

> Laboratory, radiology, and ultrasound equipment is insufficient at the secondary and tertiary levels, in spite of recent investments. Existing equipment and facilities tend to wear out and/or reach the end of their useful life, with insufficient money available to maintain and/or replace them, even in the private sector.

Nicaragua is involved in cooperative efforts, by way of the Joint Country Learning Assessment (JCLA) initiative, to align external aid with its health needs. In 2004, 24% of USD 51.3MM received by the health sector as "health cooperation" was targeted towards "strengthening medical infrastructure and equipment" (PAHO 2007).

Upon examination of the inventory of "basic equipment necessary to provide primary care" in the public sector, PAHO determined that just 75% of the 1,682 pieces of equipment in MINSA facilities are "in good operating condition" (2007, 2008). Further inquiry reveals the statistics shown in Table 3.4, which were reported about MINSA Facilities. Similar challenges are faced by the private sector, as shown in Table 3.5.

Table 3.4 Public sector equipment challenges

Equipment category	% of large health centers[a] containing equipment	% of health posts[b] containing equipment
Refrigerator	100	61
Cold box	100	28
Vaccine carriers	100	82
Communications (radios and telephones)	67	<25
Emergency transport service	67	3

[a]"Large" refers to health centers with hospital beds
[b]Health posts includes all health posts and health centers without hospital beds
Sources: PAHO (2007, 2008)

In tertiary facilities (hospitals), even the equipment that is present does not serve its intended function; of "an estimated 7,705 hospital medical equipment systems, 73% are functioning properly, 9% irregularly, and 18% [are] out of service" (PAHO 2008).

This information reveals that a stunningly high proportion of health facilities, even in the private sector, are unequipped (on the basis of available equipment) to provide primary care in accordance with the national standard. The larger facilities in the public sector seem to have the most consistently high figures, although an accurate picture of the situation would require information on what proportion of equipment are functioning properly in each sector. When advanced equipment is considered, the situation is even more extreme; the 2009 WHO medical device survey reports only one each of magnetic resonance imaging (MRI) machines, nuclear medicine devices, linear accelerators, and telecobalt units as well as three computerized tomography (CT) scanners available in the nation.

This suggests that perhaps the MINSA has greater incentive or ability to distribute resources effectively than the private sector. In any case, the data reveals the disparity in outcome between urban and rural areas, which contain the higher proportion of health posts (versus health centers). Providing a remedy to this situation will require a number of steps, including review of the sources of equipment in the private sector and coordination on the part of a single agency (potentially within the MINSA) to ensure adequate distribution of the necessary supplies. It is important to note, however, that this addresses only the problem of equipment distribution. The functionality of equipment is tied to a much deeper problem related to some of the concerns discussed in Chap. 1 (particularly in the case of temperature-dependent equipment). Availability of telecommunications and other devices that depend on local infrastructure is often less a function of a specific facility's inventory than of local underdevelopment or other problems that require systemic change to be spearheaded by one or more government departments. This argument reinforces the need for health care in developing nations to be addressed in the larger context of a broad development agenda.

Table 3.5 Private sector equipment challenges

Equipment category	% of establishments containing equipment
Refrigerator	38
Cold box	6.9
Vaccine carrier	34
Radio	9
Telephone	87
Emergency medical transport	33

Source: PAHO (2008)

Table 3.6 Leading causes of death in Nicaragua, 2010

Cause	Rank
Coronary heart disease	1
Diabetes mellitus	2
Stroke	3
Kidney disease	4
Liver disease	5
Influenza and pneumonia	6
Lung disease	7
Hypertension	8
Road traffic accidents	9
Stomach cancer	10

Source: World Health Rankings

MTS Methodology (with Calculation)

In order to demonstrate the use of the MTS using real inventory data, we can follow a similar procedure described in Chap. 2.

Step 1: Identify the top causes of local mortality/morbidity
The leading causes of death in Nicaragua (2010) are shown in Table 3.6.

Step 2: Use WHO guidelines for equipment necessary to diagnose and treat these conditions
Nicaragua's medical device nomenclature system is based on the Global Medical Device Nomenclature (GMDN) (WHO 2010c). Furthermore, the nation has developed a list of recommended medical devices for specific facilities and procedures. In a more comprehensive calculation, that information would be included here as an additional parameter.

The WHO medical device survey (2010c) for Nicaragua contains inventory information for 32 hospitals, 10 of which are specialty hospitals (Table 3.7). The equipment listed for each hospital is classified under one of the following categories: electro-medical equipment, electromechanical equipment, and refrigeration and air conditioning equipment.

The electro-medical equipment category contains diagnostic, imaging, and laboratory items that are either used in the laboratory setting or used directly to treat patients. Electromechanical equipment includes equipment such as incinerators, ovens, washing machines, and sterilization units that are used for equipment preparation, storage, or cleaning. Clinical infrastructure (such as surgical and gynecological

Table 3.7 Summary of WHO-Nicaragua inventory

	Institution name	1°/2°	General/specialty hospital?	Items[a] in inventory
1	Hospital Antonio Lenin Fonseca-*Managua*	–	**General (Teaching)**	**17**
2	Hospital Rehabilitación Aldo Chavarria-*Managua*	–	Specialty (Rehabilitation)	1
3	Hospital José Dolores Fletes-*Managua*	–	Specialty (Psychiatry)	15
4	Hospital Fernando Velez Paiz-*Managua*	–	Specialty (Maternity/Children)	38
5	Hospital Bertha Calderón-*Managua*	–	Specialty*(Women)	37
6	Centro Nacional De Dermatología-*Managua*	–	Specialty (Dermatology)	12
7	Centro Nacional De Cardiología-*Managua*	–	Specialty (Cardiology)	7
8	Hospital Alemán Nicaragüense-*Managua*	–	**General**	**41**
9	Hospital Roberto Calderón-*Managua*	–	**General**	**30**
10	Centro Nacional De Radioterapia-*Managua*	–	Specialty (Radiology)	2
11	Centro Nacional De Oftalmología-*Managua*	–	Specialty (Ophthalmology)	7
12	**Hospital Dr. Ernesto Sequeira B.-*Bluefields (RAAN)***	2	General	34
13	Hospital San José-*Carazo*	–	**General**	**13**
14	**Hospital Santiago-*Carazo***	2	General	21
15	**Hospital Cesar Amador Molina-*Matagalpa***	2	General	30
16	[Hospital Oscar Danilo Rosales-*León*]	2	General	0
17	Hospital Gaspar García Laviana-*Rivas*	2	**General**	**19**
18	Hospital Humberto Alvarado-*Masaya*	–	**General**	**33**

(continued)

Table 3.7 (continued)

	Institution name	1°/2°	General/specialty hospital?	Items[a] in inventory
19	Hospital De Amistad Japón Nicaragua-*Granada*	–	**General (Private)**	**46**
20	Hospital Victoria Motta-*Jinotega*	2	**General**	**18**
21	Hospital Alfonso Moncada-*Ocotal*	–	**General**	**24**
22	**Hospital Asunción Juigalpa-*Juigalpa***	–	General	25
23	Hospital José Nieborosky-*Boaco*	2	**General**	**19**
24	Hospital España-*Chinandega*	2	**General**	**37**
25	**Hospital Pedro Altamirano La Trinidad-*Estelí***	1	General	23
26	Hospital Rosario Lacayo-*León*	2	**General**	**5**
27	Hospital Mauricio Abdalah-*Chinandega*	2	Specialty (Maternity/Children)	22
28	Hospital Manuel De Jesus Rivera-*Managua*	–	Specialty (Infant)	35
29	[Hospital San Juan De Dios-*Estelí*]	2	General	0
30	[Hospital Nuevo Amanecer-*Puerto Cabezas (RAAS)*]	2	General	0
31	[Hospital Luis Felipe Moncada-*Río San Juan*]	2	General	0
32	[Hospital Juan Brene Palacios-*Madriz*]	–	General	0

Source: MINSA Nicaragua

Locations are provided in italics. Bold type indicates that the institution is locally identified as a regional hospital. The second column lists the MINSA-defined classification (primary or secondary) of relevant facilities. Bracketed institutions were excluded from the analysis because no inventory items were provided for them

[a]This refers to distinct items and does not reflect item quantities

tables) is also included in this category. The final category includes central air conditioning, refrigeration and freezing devices, and other temperature-controlled devices. For the purpose of this example,[1] we will use only the electro-medical equipment listed in the Nicaragua inventory.

Step 3: Combine with WHO standards for health care facilities to create an "ideal inventory"

The quantities used to construct the ideal inventories for each hospital were taken from UNICEF's planning guide for Primary Health Care Centres and First Referral Level Hospitals (2005). UNICEF refers to a primary health care center as an "ambulatory curative care" facility with a catchment area of 5,000–10,000 people. The first

---

[1] Data constraints make this category the easiest to use for the sake of comparison.

Table 3.8 Population served by Nicaragua's health facilities

Facility	Total	% of health care facilities	Density (per 100 k)	Population served[a]
Health post	890	80.47	15.4977	6,000
Health center	158	14.29	2.7513	36,000
District hospital	26	2.35	0.4527	221,000
Provincial hospital	26	2.35	0.4527	221,000
Regional hospital	6	0.54	0.1045	957,000

[a]Rounded to the nearest 1,000 people

referral level hospital is described as a "hospital at a district level," or a facility that serves 50,000–500,000 people. The densities of the five facility types in Nicaragua are reproduced in Table 3.8 for emphasis.

Based on the population served by health facilities in Nicaragua, health posts correspond to the UNICEF "primary health care center." District hospitals and provincial hospitals fall squarely in the population range for the first level referral hospital. For simplicity, we will include health centers in this category as well (they have significantly more equipment than rural health posts). Because more than 99% of Nicaraguan health facilities (according to Table 3.8) can be classified as one of the two facilities defined by UNICEF, the UNICEF planning guidelines serve as an appropriate source for the ideal inventory of the Nicaraguan hospitals under study. The MTS calculation method outlined in Chap. 2 can be applied here.

For the purpose of demonstration, a hospital that could be classified as a district/provincial hospital would serve as the optimal model. Of the 22 general (non-specialty) facilities listed in Table 3.7, 5 are classified locally as regional hospitals (MINSA), and 5 have been excluded from our analysis. The remaining 12 hospitals are somewhat diverse with regard to geographic location, private/public status, and level of inventory detail. Given the regional disparities discussed above between health facilities on the Pacific Coast and those on the less populous Caribbean Coast, a comparison of two facilities in close proximity to one another is more equitable and will yield more meaningful results. For this example, two such hospitals are considered: Hospital Alemán Nicaragüense (230 beds) and Hospital Roberto Calderon (250 beds), which are both public hospitals located in Managua (Wheeler et al. 2001; SODI 2011). They will hereafter be referred to as HAN and HRC, respectively. A more expansive version of the MTS model faces the challenge of accounting for geographic and demographic differences among a much larger and more varied set of facilities.

For the two hospitals chosen, the electro-medical portion of the inventory was first translated. Each item was assigned its GMDN number using the WHO "Medical Devices by Health Care Facility" document. Several of the devices in both inventories were not listed in the WHO document, and their GMDN codes are omitted in Table 3.9.

Table 3.9 Individual inventories for HAN and HRC

Hospital	Nombre del Equipo	Equipment name	GMDN code	Quantity	UNICEF quantity
HAN	Afilador de Cuchillo	Knife sharpener		1	
HAN	Aspirador	Aspirator (Suction)	36777	18	105
HAN	Auto Clave con vapor integrado	Integrated Steam Autoclave		2	2
HAN	Balanza Adulta c/ Tallimetro	Adult scale with height measurement		14	5
HAN	Balanza de mesa	Table scale		3	5
HAN	Balanza Neonatal	Neonatal scale		1	6
HAN	Balanza pediátrica mesa	Pediatric table scale*		1	
HAN	Baño María	Water bath		3	5
HAN	Capnografo c/Oxim.	Capnograph with oximeter		6	–
HAN	Centrifuga	Centrifuge	36465	4	4
HAN	Centro de Embebido	–		1	–
HAN	Cortador de Yeso	Cast cutter		1	–
HAN	Cortadora de Gasas	Gauze cutter		1	–
HAN	Cuna Térmica	Warmer		3	4
HAN	Desfibrilador	Defibrillator	37805	3	5
HAN	Detector Fetal Doppler	Fetal Doppler detector	35067	1	6
HAN	Electro Estimulo	Electrostimulus		5	–
HAN	Electrocardiograma Móvil	Mobile electrocardiogram	11407	1	1
HAN	Electrocauterizador	Electrocauterizer		4	–
HAN	Fotómetro	Photometer		1	2
HAN	Horno Esterilizador	Sterilizing oven		4	4
HAN	Incubadora de neonato	Neonatal incubator*	36025	3	5
HAN	Incubadora de transporte	Transport incubator		3	1
HAN	Lámpara fototerapia	Phototherapy lamp	35239	1	–
HAN	Lector Micro Hematocrito	Microhematocrit reader		1	2
HAN	Maquina de Anestesia	Anesthesia machine	44469	6	2
HAN	Microcentrifuga	Microcentrifuge		1	2
HAN	Microcospio	Microscope		4	6
HAN	Microtomo	Microtome		1	1
HAN	Monit Cardíaco-Desfibril	Cardiac monitor + defibrillator	37805	2	6
HAN	Monitor-Desfibril-ECG	Cardiac monitor/ defibrillator/ECG*	35195	8	
HAN	Monitor de Signos Vitales	Vital signs monitor		9	9
HAN	Procesador de Placa	Card processor		2	–
HAN	Pulsioxímetro	Pulse oximeter	17148	1	–
HAN	Rayos X Portátil	Portable X-ray machine*	37605	2	–
HAN	Rayos X Fijo	Stationary X-ray machine	37644	2	1
HAN	Rayos X Proy Arcor en C	C-arm X-ray machine*		1	
HAN	Rotador Serológico	Blood mixer		2	1
HAN	Taladro ortopédico	Orthopedic drill		2	–
HAN	ultrasonido	Ultrasound	40761	2	2
HAN	Ventilador Volumétrico	Ventilator (volumetric)*		1	4
HRC	Afilador de cuchillo	Knife sharpener		1	–

(continued)

Table 3.9 (continued)

Hospital	Nombre del Equipo	Equipment name	GMDN code	Quantity	UNICEF quantity
HRC	Aspirador	Aspirator (Suction)	36777	22	105
HRC	Baño de maría	Water bath		1	5
HRC	Centrifuga	Centrifuge	36465	4	4
HRC	Microtomo	Microtome		2	1
HRC	Desfibrilador	Defibrillator	37805	5	5
HRC	Pichel térmico	Thermal mug		1	–
HRC	Plato caliente	Hot plate		1	3
HRC	Electrocauterio	Electrocauterizer		4	–
HRC	Electrocardiógrafo	Electrocardiograph	11407	2	1
HRC	Incubadora	Incubator		2	5
HRC	Rotador de frasco	Flask rotator (shaker)		1	1
HRC	Maquina de anestesia	Anesthesia machine	44469	5	2
HRC	Microscopio	Microscope	35484	3	6
HRC	Monitor EKG	EKG monitor	35195	1	6
HRC	Procesadora de Placa Aut.	Automatic card processor		1	–
HRC	Rayos x móvil	Portable X-ray machine	37605	1	–
HRC	Rayos x fijo	Stationary X-ray machine	37644	2	1
HRC	Ventilador	Ventilator		4	4
HRC	Monitor cardiaco	Cardiac monitor	13514	8	6
HRC	Oximetro de pulso	Pulse oximeter	17148	7	–
HRC	Contador de célula	Cell counter	35103	1	3
HRC	Contador de colonia	Colony counter*		1	
HRC	Diagnostico 500	–		1	–
HRC	Horno secador	Drying oven		1	3
HRC	Lector de hematocrito	Microhematocrit reader		1	2
HRC	Mezclador de sangre	Blood mixer		1	1
HRC	Mezclador de tubo	Test tube mixer (vortex)		1	2
HRC	Taladro ortopedia	Orthopedic drill		1	–
HRC	Ultrasonido	Ultrasound	40761	2	2

Source: WHO Baseline Country Survey on Medical Devices-Nicaragua
Starred items were included in a more general category in the UNICEF planning document. They are merged with the larger category in Table 3.10

An "ideal inventory" was then constructed using the ideal quantities of each device given in the UNICEF planning guide. In order to compare both hospitals against the same baseline inventory, the ideal inventory contains the combination (minus duplicates) of "UNICEF quantities" for the two devices. Equipment that was not listed in the UNICEF planning guide is displayed in Table 3.9 but was excluded from the MTS calculation process. The ideal inventory is shown in Table 3.10.

Table 3.10 Ideal inventory for HAN and HRC

Nombre del Equipo	Equipment name	GMDN code	Ideal quantities
Aspirador	Aspirator (Suction)	36777	105
Auto Clave con vapor integrado	Integrated Steam Autoclave		2
Balanza Adulta c/Tallimetro	Adult scale with height measurement		5
Balanza de mesa	Table scale		5
Balanza Neonatal	Neonatal scale		6
Balanza pediátrica mesa	Pediatric table scale*		
Baño maría	Water bath		5
Centrifuga	Centrifuge	36465	4
Contador de célula	Cell counter	35103	3
Contador de colonia	Colony counter*		
Cuna Térmica	Warmer		4
Desfibrilador	Defibrillator	37805	5
Detector Fetal Doppler	Fetal Doppler detector	35067	6
Electrocardiograma Móvil	Mobile electrocardiogram	11407	1
Fotómetro	Photometer		2
Horno Esterilizador	Sterilizing oven		4
Horno secador	Drying oven		3
Incubadora	Incubator		5
Incubadora de neonato	Neonatal incubator	36025	5
Incubadora de transporte	Transport incubator		1
Lector Micro Hematocrito	Microhematocrit reader		2
Lector de hematocrito	Microhematocrit reader		2
Maquina de Anestesia	Anesthesia machine	44469	2
Mezclador de sangre	Blood mixer		1
Mezclador de tubo	Test tube mixer (vortex)		2
Microcentrifuga	Microcentrifuge		2
Microcospio	Microscope		6
Microtomo	Microtome		1
Monitor Cardíaco-Desfibril	Cardiac monitor + defibrillator	37805	6
Monitor-Desfibril-ECG	Cardiac monitor/defibrillator/ECG*	35195	
Monitor cardiaco	Cardiac monitor	13514	6
Monitor de Signos Vitales	Vital signs monitor		9
Monitor EKG	EKG monitor	35195	6
Plato caliente	Hot plate		3
Rayos X Fijo	Stationary X-ray machine	37644	1
Rayos X Proy Arcor en C	C-arm X-ray machine*		
Rotador de frasco	Flask rotator (shaker)		1
Rotador Serológico	Blood mixer		1
Ultrasonido	Ultrasound	40761	2
Ventilador	Ventilator		4
Ventilador Volumétrico	Ventilator (volumetric)*		4

Starred items were included in a more general category in the UNICEF planning document. For this example, they were grouped with the item in the preceding row

Step 4: Compare national inventory to ideal inventory to obtain the compliance ratio

In Chap. 2, the "item score" for item *i* was defined as

$$score_i = \frac{x_i}{y_i}, \text{where } i = A, B, C, D, E \tag{3.1}$$

Consequently, the hospital score for a facility containing five items (A through E) is

$$score_{hosp} = \frac{\sum_{i=A,B,C,D,E} (score_i)(e_i)}{n} \tag{3.2}$$

This method was applied to HAN and HRC, resulting in a score for each hospital that reflects both level of adherence of each hospital to the ideal inventory and essentiality[2] of each item (Note: The facility score in equation 3.2 is divided by the number of items (n) to equalize small and large facilities because their inventory sizes are likely to be substantially different. This step is excluded for this example.) The item scores and essentiality scores for HAN and HRC are summarized in Table 3.11.

Step 5: Weight by "essentiality" score of inventory items to create the facility score. Weight average facility scores of different facility categories by density of facilities (WHO 2010c) to create a dimensionless national "score"

Taking the average of these item scores, weighted by the essentiality of each item, results in the following values of $score_{hosp}$:

	HAN	HRC
$score_{hosp}$	0.5621	0.6065

HRC's score is slightly higher than that of HAN. Care should be taken when interpreting this metric; it is inaccurate to state that HRC has a "higher level of technology" than HAN, particularly in light of the many factors that are captured by this number. Given a larger number of hospitals in the same category, it would be possible to obtain a category score ($score_c$) for Nicaraguan district hospitals using (3.3):

$$score_c = \frac{\sum_{i=1}^{N_c} score_{hosp,i}}{N_c}, \tag{3.3}$$

Systematic application of this method results in the generation of a useful metric that can inform policy at the regional, national, and international levels. The simplest

---

[2] The procedure for determining essentiality is described in detail in Appendix C.

Table 3.11 HAN/HRC item scores and essentiality scores

Equipment name	HAN item scores	HRC item scores	e
Aspirator (Suction)	0.17	0.21	0.4
Integrated Steam Autoclave	1	0	0.5
Adult scale with height measurement	2.8	0	0.25
Neonatal scale	0.17	0	0.25
Table scale	0.8	0	0.25
Water bath	0.6	0.2	0.3
Centrifuge	1	1	0.45
Cell counter	0	0.67	0.3
Warmer	0.75	0	0.4
Defibrillator	0.6	1	0.85
Fetal Doppler detector	0.17	0	0.6
Mobile electrocardiogram	1	0	0.65
Photometer	0.5	0	0.45
Sterilizing oven	1	0	0.45
Drying oven	0	0.33	0.4
Incubator	0	0.4	0.7
Neonatal incubator	0.6	0	0.7
Transport incubator	3	0	0.75
Microhematocrit reader	0.5	0	0.5
Hematocrit reader	0	0.5	0.5
Anesthesia machine	3	2.5	0.95
Blood mixer	2	1	0.45
Test tube mixer (vortex)	0	0.5	0.45
Microcentrifuge	0.5	0	0.45
Microscope	0.67	0.5	0.55
Microtome	1	2	0.5
Cardiac monitor + defibrillator	0.33	0	0.65
Cardiac monitor	0	1.33	0.6
Vital signs monitor	1	0	0.55
EKG monitor	0	0.17	0.65
Hot plate	0	0.33	0.35
Stationary X-ray machine	3	2	0.65
Flask rotator (shaker)	0	1	0.45
Blood mixer	2	1	0.45
Ultrasound	1	1	0.55
Ventilator	0.25	1	0.9

interpretation that can be drawn from a comparison of MTS scores is a relatively higher or lower level of adherence to the *specific* ideal inventory that corresponds to a given hospital.

A second goal of the MTS is to capture a given facility's ability to treat conditions that are locally relevant. In other words, it is useful to know whether a hospital is equipped to treat the diseases that factor significantly into the burden of disease in the surrounding area. In constructing a quantitative measure of disease treatment capability, documents such as the WHO Availability Matrix (Hansen et al. 2010) can serve a role similar to the documents used in this example.[3]

The preceding example is only an initial expression of the MTS score as conceptualized by the author; much refining is still needed before this metric can be utilized as proposed. This is an on-going task for the proponent of this score. Several challenges were met during the construction of the MTS for Nicaraguan hospitals. Lack of standardization among the reference documents used to create the baseline inventory is a major obstacle that will need to be overcome in order for widespread use of this method to yield actionable results.

Furthermore, several assumptions were made regarding the methods of data collection and completeness of the inventories used. Ministries of Health wishing to change technology policy in a manner that reflects the true situation of national health facilities must standardize and monitor the regular collection of inventory data across institutions. At the point when the development of the score has been completed and fine tuned to the point of effective usage by both donor and recipient countries, coordination of these efforts across countries will be best accomplished with the support of international organizations such as UNICEF and the WHO.

Agriculture

Nicaragua's agricultural output is currently insufficient to meet the food needs of the nation. Unlike Ghana, "the productive infrastructure [in Nicaragua] tends to be concentrated in urban areas…the agricultural sector contributes less than 20% of GDP" (PAHO 2008).

In Nicaragua, agricultural income at the household level is much less biased toward crop production than in Ghana, where crop production is the predominant method of generating agricultural income. On average, the raising of livestock is the single largest contributor to household income (Fig. 3.5).

In 2009, the top five crops in Nicaragua (as measured by production) were sugar cane, whole milk, maize, rice, and beans (FAOSTAT). Detailed breakdowns of the local and regional production of maize and soy by Nicaragua's neighbors and other member countries of Central America are shown in Tables 3.12 and 3.13, respectively.

Although the production of soy in Central America is only 1% that of maize production, Nicaragua is located within close proximity of the United States, which is the world's top producer and exporter of soy. In 2009, the United States produced 91,417,300 tons and exported 40,505,700 tons of soybeans. If we include this figure,

---

[3] UNICEF PHC planning guide and WHO facility-based inventory.

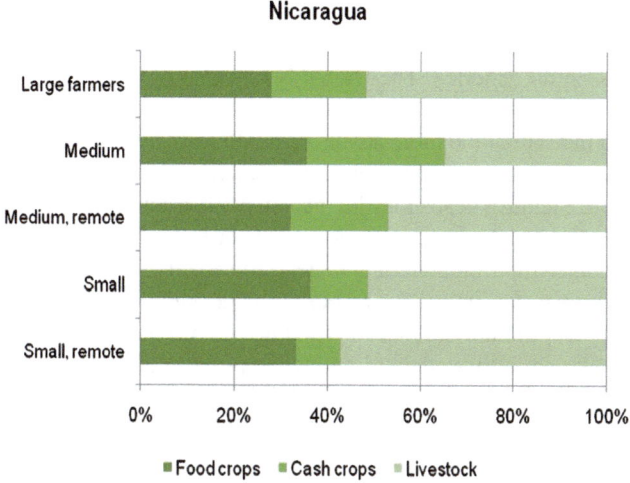

Fig. 3.5 Farm household production mix — Nicaragua (Brooks et al. 2011)

Table 3.12 Maize and soybean production by Nicaragua and surrounding countries

Item	Country	2009 Production (tons)
Maize	Costa Rica	24,005
	Honduras	5872,35
	Nicaragua	522,024
Soybeans	Costa Rica	
	Honduras	1,868
	Nicaragua	2,300

Source: FAOSTAT

Table 3.13 Maize and soybean production in Central America

Maize		Soybean	
Country	2009 Production (tons)	Country	2009 Production (tons)
Belize	45,041	Belize	585
Costa Rica	24,005	Costa Rica	
El Salvador	785,965	El Salvador	3,905
Guatemala	1,686,890	Guatemala	37,000
Honduras	587,235	Honduras	1,868
Nicaragua	522,024	Nicaragua	2,300
Panama	85,544	Panama	89
Total	3,736,704	Total	45,747

Source: FAOSTAT

the total quantity of soybeans (theoretically) locally available to the Nicaraguan agricultural sector is 91,463,047 tons—more than 24 times the quantity of maize produced in Central America. The United States also produces 47,813,400 tons of maize, or 12 times the maize output of the Central American region (FAOSTAT).

Nicaragua has one of the more advanced health care systems in Central America and may be an excellent candidate for a biomaterials industry. While its own production of maize and soybeans may be inadequate for such an industry, Nicaragua's positioning in Central America provides it the access to both crops that it needs to become the center for a biomaterials industry. In reality there would have to be an intentional effort to tap into these resources, especially given that the nation currently imports crops simply to meet its food needs. A separate batch of crops would likely need to be pre-allocated and imported explicitly to supply an in-country biomaterials industry. Depending on the feasibility of these changes, it may be more practical to consider the creation of a regional industry that serves all of Central America. The OECD supports this view, speculating that cooperation among Mesoamerican (Central America plus Mexico) nations could result in a region that optimally takes advantage of trading giants such as the United States and collectively "offers year-round primary agricultural production of a variety of high-quality traditional and non-traditional products, as well as processing high value-added agro-industrial products" (OECD 2006).

Discussion and Conclusion

This chapter has provided an overview of the structure of the Nicaraguan health and agricultural sectors, a brief political history as it relates to the current health situation, and a discussion of health care technology and the burden of disease in Nicaragua and the surrounding regions of Central/Mesoamerica.

Nicaragua has made an epidemiological transition; while several infectious, tropical, and communicable diseases are still formidable challenges to public health, chronic and incommunicable diseases have emerged as the newest frontier in Nicaragua's disease landscape. The prevalence of formerly focal illnesses such as malaria has declined over time. In addition, the renewed commitment of the government to nationwide vaccination coverage has resulted in reduced prevalence of vaccine-preventable illnesses. The effectiveness of the "National Health Day" vaccination campaign has been one of the hallmarks of the new Nicaraguan health system, as has incorporation of a social element into a new system skewed toward "popular" participation and decentralization.

Health care equipment in Nicaraguan health facilities at all levels is inadequate. Many facilities have insufficient equipment, and the equipment in place is often functioning poorly, if at all. Accurate determination of the source of this problem requires that we identify the source of the equipment. Discussions with health care personnel suggest that a significant amount of this equipment is new or secondhand equipment donated from more industrialized nations such as the United States. It is unclear how much of this equipment transfer is regulated, as Nicaragua has no national medical technology policy (WHO 2010c). However, PAHO et al. (2001) have developed a model regulatory program for medical devices. A policy priority over the coming decades should be the development of more stringent and context-specific guidelines. This will have the dual effect of ensuring that all imported or donated equipment meets a certain standard of excellence (including appropriateness for the given context, especially in rural areas) and promoting local generation of alternative equipment to take the place of equipment that is no longer deemed acceptable.

An attempt was made to calculate a MTS for two of Nicaragua's hospitals. The MTS is still in the preliminary stages of development, but it has the potential to convey the abilities and needs of facilities within a country. In the future, it may serve as an additional tool for regulation of equipment donation. The MTS will also serve to convey the potential capacity of the facility to provide certain services that have advanced technological requirements.

Agricultural income in Nicaragua comes from a relatively wide range of activities, the largest of which is the raising of livestock. The OECD Territorial Review on Mesoamerica (2006) specified that

> Labour income is the main factor in determining family income in Mesoamerica, yet the labour market is filled with low-productivity, low-wage and unstable jobs that often do not cover health and insurance benefits. Since the majority of new jobs are created in the informal sector, the productivity and income gains for the region are limited.

This indicates that a renewed focus on crop production and other income-generating activities may actually benefit the region. However, the constrained resources and mismatch of agricultural output to food needs suggests that Nicaragua may be at its capacity in some respects. A prudent strategy for the future must take this into account and look outward to explore how Nicaragua can make use of partnerships with other Central American Nations and use trade to its advantage.

Nicaragua is a potential location for a biomaterials industry in Central America because of its leadership in health care reform among Central American countries. Because of its unique geographic location, Nicaragua has abundant access to both maize and soybean, as well as a number of other crops. The limiting factors to the generation of a biomaterials industry are meeting the challenges of administration in an increasingly decentralized health system and coordinating among the numerous sectors that would have a stake in such an industry. These are all challenges to ponder as the Nicaraguan health system comes of age.

Chapter 4
Corn- and Soy-Derived Materials: Properties and Potential Clinical Applications

Biomaterials

This discussion of biomaterials is focused on implantable materials that "comprise whole or part of a living structure or biomedical device which performs, augments, or replaces a natural function" (Park 1979). Chapter 1 refers to the first category described above—the materials of biological origin—as "natural materials," which have the potential to be implemented as biomaterials. "Syntheto-naturals" can then be defined as the set of materials that are derived in the natural (most likely agricultural) setting but that can be manipulated and adapted for use as biomaterials, like many synthetic implantable materials.

An important consideration for use of biomaterials in (and on the surface of) implants is the expected response of host cells. Ratner and Bryant (2004) thoroughly characterize the components of the host response to implantation:

1. Nonspecific protein adsorption of cells that are normally key players in wound healing.
2. Upregulation of cytokines and subsequent pro-inflammatory processes.
3. Unsuccessful attempt of adhered macrophages to phagocytose the foreign body.
4. Chronic inflammation at the implant interface.
5. Fusion of macrophages to form multinucleate foreign body giant cells.
6. Walling off of the device by an avascular, collagenous fibrous tissue (can cause complications).

Because we cannot predict the biological response to implanted biomaterials, clinicians currently respond by treating after the fact. One goal in the design of novel biomaterials for use in tissue engineering and medicine is the ability to control the host response via surface chemistry and architecture modification (Ratner and Bryant 2004). In the meantime, the evidence provided by the study of previously implanted materials serves as a guide. The *in vitro* response to implants rarely mirrors the results achieved in the laboratory, presenting an additional

O.A. Fatunde and S.K. Bhatia, *Medical Devices and Biomaterials for the Developing World: Case Studies in Ghana and Nicaragua*, SpringerBriefs in Public Health, DOI 10.1007/978-1-4614-4759-7_4, © Springer Science+Business Media New York 2012

challenge for the design of implantable biomaterials. Important parameters include porosity, which has implications on the vascularization of the area surrounding the implant, stiffness, and strength of the material. Levental et al. (2007) explored the production of hydrogels (alone or combined with natural biopolymers) that mimic or possibly control natural conditions.

Clinical Applications of Biomaterials

The health issues to be considered draw from two sources. The first will depend on the results of our biocompatibility studies (for example, if we were to discover that corn- and soy-derived materials are only compatible with cardiac fibroblasts, we would focus on clinical applications that can impact treatment of heart conditions). The second is the burden of disease of the countries under study, which will allow us to draw relevant conclusions. The most highly prioritized conditions will be those that fall at the intersection of the two.

Bhatia (2010) reviews potential application of biomaterials for various clinical applications. What follows is a review[1] of these applications as they relate to several diseases. Of particular interest are the following:

Cardiac Disease

Coronary artery disease results from atherosclerosis, or the gradual buildup of plaque on the inner lining of blood vessels. The causes are numerous and draw from a combination of genetic inheritance, lifestyle (including diet), and living environment. The immediate cause of atherosclerosis is the buildup of cholesterol and various immune cells. Coronary artery disease thus "represents the culmination of cholesterol accumulation, cellular capture, vascular injury, and inflammatory activation" (Bhatia 2010, 27). Arterial blockage can result in myocardial infarctions, or in extreme cases, even heart failure.

Biomaterials represent one option for treatment of both atherosclerosis and heart failure. Blocked coronary arteries have traditionally been opened by devices called stents, which are cylindrical wire meshes that can restore the compromised vessel to a healthy diameter. Cells eventually populate the stents, which become permanently embedded in arterial walls. Although stents are mechanically expanded upon implantation, they do allow for recurrence of stenosis over time. Re-stenosis is due

---

[1] Page numbers from Bhatia (2010) will be included in the citations for this section.

to both the damage that the stent does to the artery wall and the tendency for cells to over-proliferate on the stent surface.

The use of biomaterials to create bioactive stents can reduce the risk of re-stenosis, provided that the biomaterials to be used are sufficiently flexible, compatible with the biological environment of the bloodstream, and sterilizable (Bhatia 2010, 34). Stents can also be altered so that they are drug-eluting, providing pharmacological resistance to over-proliferation of cells, or antibody-coated, preventing the artery from re-narrowing.

Another possible application is the creation of degradable stents using biomaterials. Degradable stents are a novel solution to three problems commonly encountered with the use of metal stents. Metal stents constrict blood vessel and force them to reshape themselves around the diameter of the stent, which cannot increase. They also trigger long-term clotting and make subsequent surgeries, such as implantation of additional stents, quite difficult (Bhatia 2010, 39).

The replacement of metal stents with those made of biomaterial polymers is difficult due to the need to maintain strength without losing the appropriate flexibility/ rigidity balance. Polymers of the PLA family, one of which will be discussed later in this chapter, are under consideration as a candidate material for biodegradable stents. Copolymers involving the two optical isomers of PLA (PDLA and PLLA) have been identified as possibilities for use in both biodegradable and drug-eluting stents.

There is a real opportunity for biomaterials to fill the need for effective heart failure treatments. Due to the significant loss of cells that occurs during a heart attack, a tissue engineering approach to congestive heart failure treatment involves re-growth of cardiac myocytes on a scaffold in order to replace lost cells and restore functionality to the damaged tissue. The scaffolding material should be degradable, elastic, and electrically stable. It should also be able to support cell proliferation and differentiation. There are two methods of cardiac regeneration via tissue engineering: the *in vitro* method, which involves implantation of a scaffold with cells already growing on it, and the *in situ* method, which requires direct injection of cells into the affected tissue. Progress has been made towards clinical use of both methods (Bhatia 2010, 45), but the search for additional materials with the desirable properties continues in the hopes of increasing survival rates for one of the world's most deadly diseases.

Therapy After Traumatic Injury

Road traffic accidents and other traumatic injuries claim lives daily all over the world. Unlike many of the conditions described in this section, traumatic injury is a result not of infection or sudden dysfunction of body tissues due to disease, but rather the interruption of one or more body functions due to an unexpected external event.

Table 4.1 Classes of biomaterials and their possible applications for healing traumatic injuries

		Application for traumatic injury
Biomaterial		
1	Fibrin-based tissue adhesives	Wound closure (principal biological sealant)
2	Cyanoacrylate-based tissue adhesives	Tissue glue (principal synthetic polymer sealant)
3	Cross-linked protein-based adhesives	Tissue glue (less toxic than (2))
4	Polyethylene Glycol (PEG)-base adhesives	Wound closure (synthetic hydrogels)
Emerging biomaterials		
5	Naturally inspired tissue adhesive	Tissue glue (naturally occurring in frogs, mussels)
6	Polysaccharide-based tissue adhesives	Wound closure
7	Dendrimeric tissue adhesives	Tissue glue (hydrogel sealant)

Source: "Road Traffic Accidents," Bhatia (2010)

Treatment following traumatic injury is heavily dependent on the level of trauma and the body part affected, but in many cases surgery is required to close open wounds or heal tissues that have been severed or otherwise disrupted.

There are numerous applications of biomaterials, particularly those with adhesive properties, in tissue reconstruction and wound closure. Current approaches to wound closure, many of which are based on suturing, result in relatively high rates of leakage, which can result in infection and increased mortality. A significant challenge exists in finding appropriate alternatives, which must have the requisite physical/mechanical properties, while also demonstrating appropriate biocompatibility and degradability (Bhatia 2010, 214). In order to justify their adoption, they must also improve on current methods. Table 4.1 lists types of materials under development and their specific applications for trauma therapy.

Many of these materials have been tested by surgeons in animal models and achieved astonishing results. Surgeons have cited ease of use as one of their top criteria for implementing these techniques in human patients. With continued discovery of new materials, and in the wake of a movement of surgery towards increasingly minimally invasive procedures, it is within the realm of possibility for biomaterial adhesives to soon be widely used alongside their older counterparts (sutures) for closing wounds and healing damaged tissue. Their adoption in the developing world brings additional challenges.

Diarrheal Disease

Diarrheal diseases are unique among the conditions discussed here because of their disproportionate impact on the poorest and youngest populations. According to Keusch et al. (2006), approximately 90% of diarrheal disease cases in the developing world take place in children under 5 years of age. Diarrhea can be caused by the

Table 4.2 ASSURED model: important characteristics of low-cost diagnostic devices

Criterion	Note
*A*ffordable	By those at risk for infection
*S*ensitive	Few false negatives
*S*pecific	Few false positives
*U*ser-friendly	Simple to perform with minimal training necessary
*R*apid and robust	To enable treatment at first visit (rapid) and not require refrigerated storage (robust)
*E*quipment free	
*D*elivered	To those in need

Source: Linnes (2011)

introduction of parasites, bacteria, or viruses into the gastrointestinal tract, and it is shared between people via fecal-to-oral transmission. As such, areas that suffer from low sanitation levels and other public health concerns tend to have a high prevalence of diarrhea.

Because the disease prevents re-absorption of ingested fluids, one of the ways that diarrhea proves to be fatal is through dehydration of sufferers. In addition to this acute watery diarrhea, sufferers can experience bloody diarrhea (resulting from intestinal damage) and persistent diarrhea, lasting 14 days or longer. Appropriate treatment, which can include rehydration, nutritional boosting, and antimicrobial therapy, depends on the type of diarrhea involved (Bhatia 2010, 131).

Because of its connection to poverty and nutrition, this disease has been a priority not only for health professionals but also for companies and NGOs with an interest in poverty alleviation. Bangladesh-based BRAC, the world's largest development NGO, was instrumental in the mass distribution of the Oral Rehydration Therapy (ORT) kits developed for diarrhea by Richard Cash in the late 1970s. The Oral Therapy Extension Program in Bangladesh lasted for 10 years, and as of 2006 ORT is estimated to have saved 40–50 million lives in the intervening years (Mushtaque et al. 1996; Fontaine et al. 2007). Although the symptoms are well defined and cost-effective treatment methods continue to be developed by organizations like BRAC, a significant challenge remains in the area of diagnosis. Definitive diagnosis of diarrhea requires extensive laboratory testing and stool culture, which are simply not feasible on a large scale in many low-resource settings. Biomaterials and innovative diagnostics present a possible solution to this problem.

Diagnostic devices, particularly those designed for use in the developing world, must be appropriate along a number of parameters (see Table 4.2). Combination of proven laboratory techniques with the use of sustainable biomaterials sufficiently addresses the scientific needs of diarrheal disease diagnosis. The prohibitive aspects of traditional models are the expensive and inaccessible infrastructure that is required to maintain permanent labs as well as the high capacity that would be required of these facilities in order for them to address the ever-increasing number of cases. In light of this, BRAC's precedent of producing a pre-prepared and -packaged treatment kit in large quantities can serve as our guide. The design of highly versatile microfluidic devices (Fig. 4.1), or small mini-reactors that can be pre-loaded

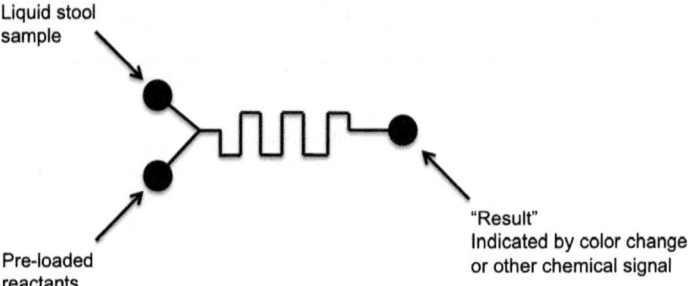

Liquid stool
sample

"Result"
Indicated by color change
or other chemical signal

Pre-loaded
reactants

Fig. 4.1 Basic microfluidic device

with small amounts of reagents, provides a diagnostic counterpart to the example of ORT. Several inquiries have been made into what makes microfluidics appropriate for the developing world (Chin et al. 2011; Martinez et al. 2010; Yager et al. 2006). Widespread adoption of microfluidic devices, and possibly other rapid paper diagnostics, would allow for almost instantaneous diagnosis and might change the landscape of treatment for diarrheal disease.

Tuberculosis

Tuberculosis (TB) is an infectious bacterial disease that kills millions annually, with the overwhelming burden of disease on the poor. Tuberculosis is caused by infection with the bacterium *Myobacterium tuberculosis*. TB transmission is airborne. Nearly 5% of all tuberculosis episodes are multidrug-resistant tuberculosis (MDR-TB). Treatment of tuberculosis in both the susceptible and drug-resistant forms lasts for an extended period (6 months or more). In the early 1990s, the WHO implemented a standard tuberculosis treatment protocol called Directly Observable Treatment, Short Course (DOTS) (Table 4.3), which drastically reduced tuberculosis-related mortality (Program in Infectious Disease and Social Change 1999).

The length of tuberculosis treatment, and the related problem of patient noncompliance with treatment protocol, is a major cause of both drug-resistant TB and TB-related mortality. Biomaterials represent a mutually advantageous solution for clinicians and TB patients. Drug delivery using biocompatible and biodegradable biomaterials is a possible replacement to the current method of TB model, which requires multiple patient visits (which can be infeasible for remote patients) or regular outreach visits by physicians or community health workers. These materials can be administered orally or directly to the lungs (Bhatia 2010, 178).

By providing an option for long-term, sustained drug release, novel biomaterial technologies reduce the burden of compliance that currently rests on the patient, allow doctors to see patients less frequently (possibly increasing efficiency), and result in improved clinical outcomes due to both direct administration and strict adherence to treatment.

Table 4.3 Components of WHO DOTS treatment regimen

Component	Explanation
1. Political commitment with increased and sustained financing	Legislation, planning, human resources, management, training
2. Case detection through quality-assured bacteriology	Strengthening TB laboratories, drug resistance surveillance
3. Standardized treatment with supervision and patient support	TB treatment and programme management guidelines, International Standards of TB Care (ISTC), PPM, Practical Approach to Lung Health (PAL), community-patient involvement
4. An effective drug supply and management system	Availability of TB drugs, TB drug management, Global Drug Facility (GDF), Green Light Committee (GLC)
5. Monitoring and evaluation system and impact measurement	TB recording and reporting systems, Global TB Control Report, data and country profiles, TB planning and budgeting tool, WHO epidemiology and surveillance online training

Source: WHO (2006)

Pneumonia

Pneumonia is a lower respiratory disease that is one of the world's leading illnesses and killers. Significant pneumonia-related child mortality plagues the developing world. Interestingly, pneumonia is as much of a problem for adults in the developed world as in the developing world, but the median age of sufferers is much higher in the developed world (Scott et al. 2000). Pneumonia is a common killer of patients who are coinfected with autoimmune diseases.

Pneumonia results in the alveolar sacs of the lung being filled with fluid, eventually leading to collapse and inability of these bodies to exchange oxygen with the bloodstream. In extreme cases, it can also lead to fluid filling the cavity that houses the lung. Pneumonia is often caused by bacteria, which can aggregate to form tough-to-penetrate biofilms, or undesirable conglomerates of bacteria adhering to a surface, in the lung cavity. As such, antibiotics are required to overwhelm the bacteria that introduce this devastating disease. However, resistance to antibiotics remains a major challenge for treatment. There is also a viral form of pneumonia that is unresponsive to antibiotics and whose treatment consists of antiviral drugs (when possible) and is largely supportive, depending on the symptoms present.

Biomaterials can be used to fight pneumonia by addressing the challenging issue of penetrating biofilms. Liposomes, spherical "colloidal biomaterials that have been designed for controlled release of pharmaceuticals," are being explored as a potential carrier for one or more antibiotic at a time. The liposome–drug combination can be administered via intravenous injection. While they have yet to be tested in humans, liposomes have improved survival rates in rats with compromised immune systems and those infected with pneumonia caused by drug-resistant bacteria (Bhatia 2010, 92). This success bodes well for the potential efficacy of this novel biomaterial.

Stroke

Stroke is a neurological condition that, along with heart attacks, kills 17 million around the world each year (WHO 2012). Often marked by sudden onset, strokes result from a failure of blood, and therefore oxygen, to reach one or more parts of the brain. Stroke can result from either blockage (ischemic) or rupture and leakage (hemorrhagic) of a blood vessel in the brain. The condition is diagnosed using advanced imaging techniques such as MRI and CT scans.

Stroke is closely related with cardiovascular disease, and as such many applications of biomaterials to treating heart disease and heart failure (described above) have the potential to help stroke patients as well. While ischemic stroke is much more common (70% of all strokes), hemorrhagic stroke is quite devastating and can result in hematoma, or pooling of blood within the brain. The two are not to be confused, particularly at the diagnostic stage, as treatment for the former can prove fatal when administered to a patient who actually suffers from the latter (Bhatia 2010, 62).

Biomaterials have the potential to decrease mortality for stroke patients both by improving the accuracy and speed of the diagnostic process (via pre-MRI labeling of macrophages using bioactive nanoparticles) and by contributing to the re-growth of neurons in order to restore lost tissue functionality. Biomaterials recruited to do the latter must meet many of the criteria discussed above. In addition, they must have the unique quality of being electrically conductive. Nanoscale materials, such as nanotubes and nanofibers, are uniquely suited to this task because of their high surface-to-volume ratios and ability to mimic native conditions. Soliman et al. (2012) conducted preliminary experiments exploring the electrical properties of carbon nanotubes and the feasibility of growing neural cells on a carbon nanotube-hydrogel matrix.

Each of the conditions described above has potential for improved diagnosis, treatment, or scaling through the use of biomaterials as a replacement for or complement to current practices. Several techniques, such as drug delivery using nanoparticles and implantation of scaffolding for cell regeneration, have applications for several diseases. The versatility and potential benefits of biomaterials, particularly sustainable or reusable materials, point to new possibilities for the future of clinically relevant engineering.

Like other medial devices, the adoption of biomaterials for clinical use in most countries requires extensive government and industry approval. In order to be deemed safe for clinical (or commercial) use, a biomaterial but be demonstrably non-toxic, effective, sterilizable, and biocompatible (Ikada and Tsuji 2000). Given the goal of improving clinical outcomes through the use of biomaterials, testing for toxicity and biocompatibility is an appropriate first step in the evaluation of candidate materials.

Preparation of Selected Biomaterials

Corn-Derived PLLA

Lactic acid is a naturally occurring compound that is produced from sugars such as glucose via fermentation of pyruvate. One of its optical isomers, L-lactic acid, is commonly found in the human body and in plants, animals, and microorganisms (Galactic 2012). One common source of lactic acid, and the one considered in this study, is corn starch. The most notable component of starch-based materials is amylose, which contributes to both the flexibility and the digestibility of the biomaterial. The starch/amylose content of a given material also has an effect on the tensile strength of the finished product (Kim et al. 1998). After lactic acid is isolated in one form or a mixture of its isomeric forms, it can be polymerized in the laboratory via a ring-opening mechanism (Fig. 4.2). Stannous octoate is used as a catalyst for this process, and lauryl alcohol is used to determine the length of the polymer. The molecular weight of the polymer is determined by the relative concentrations of these two reagents (Ikada and Tsuji 2000).

In its polymerized form, L-lactic acid is referred to as poly(L-lactic acid), or PLLA. Like other forms of PLA, PLLA is an aliphatic thermoplastic polymer, which means that it softens reversibly when heated and hardens when cooled (Balasubramaniam et al. 2007). Polymers such as starch (at a price of $1.50/kg) can be mixed with synthetic polyesters ($4/kg) to produce a blended alternative at a much lower cost. In a given unit, almost 50% of the polyester may be replaced by a polymer without sacrificing quality (Malhotra et al. 2008). PLLA can be processed into various forms such as fiber and film. Mikos et al. (1994) demonstrated that foams of varying porosity can be created using PLLA–salt composites of varying salt concentrations, which are molded and then purified via dissolution of the salt. The versatility, potential porosity, and biodegradability of PLLA signal its potential as a material with clinical applications. PLLA is also a renewable biomaterial; the fiber can be recycled via de-polymerization to L-lactic acid and re-processing (back to PLLA) after purification.

While PLLA has been identified as a material of interest, its current usage is limited to cosmetic fillers and surgical elements that either do not have direct contact with the body or are intended to degrade on a relatively short time scale, such as sutures (Goldman 2011). A possible constraint in the use of PLLA-based materials is the toxicity of the by-products of degradation, especially when the implant is present in the body for a long time such as may be the case for an implanted biomaterial. These by-products include remnants from the polymerization process and low-molecular-weight leachables, or materials that separate from the surface of the material and/or enter the environment of the cells (Ikada and Tsuji 2000). Elimination of the potentially dangerous degradation phase may mitigate the risk of infection and provide another option for implants. Assuming biocompatibility and given the many benefits of the material, a goal for future research may well be adaptation of PLLA, or other naturally derived materials, in one of its many forms for use in long-term grafts or implants.

The PLLA used in this experiment was obtained from Biomer and processed in the laboratory of Dr. Manjusri Misra and Dr. Amar Mohanty at the University of Guelph.

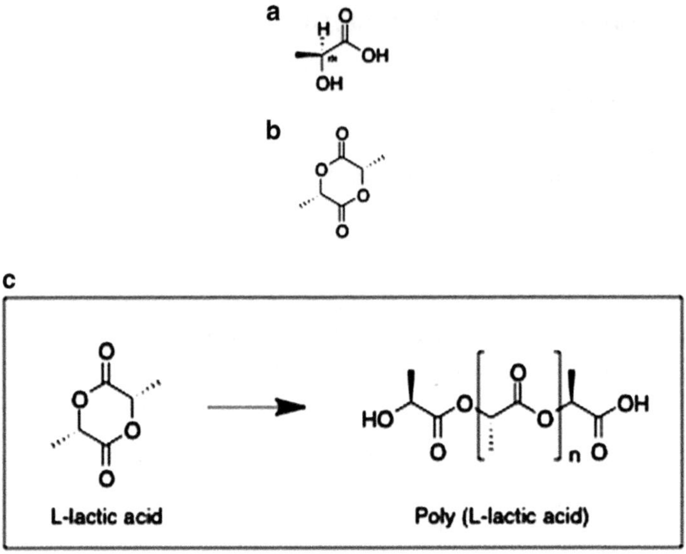

Fig. 4.2 L-Lactic acid polymerization reaction. L-lactic acid in its (**a**) single molecule form, (**b**) cyclic dilactate ester form, and (**c**) polymerized form

The solvent used to dissolve varying percentages (by weight) of PLLA was a mixture of chloroform and *N,N*-dimethylformamide (DMF). Both substances were purchased from Fisher Scientific.

The corn-derived PLLA fibers are biodegradable and composed of synthetic nonwoven material that was processed via the electrospinning[2] technique using the MECC NANON electrospinning machine. Synthetic nonwoven materials have been used in scaffolds for tissue engineering and tissue generation, and additional potential applications include protein purification and bioproduct toxin removal (Misra et al. 2011).This method of preparation has the benefit of producing nanocomposite fibers that have a high surface-to-volume ratio, high porosity, and excellent mechanical properties (Cui et al. 2010). The critical components of the electrospinning setup are pictured in Fig. 4.3. A scanning electron microscope image of PLA fibers is provided in Fig. 4.4, and important parameters are listed in Table 4.4.

The final diameter and morphology of the fibers is dependent on the properties of the solution in the syringe (Cui et al. 2010). The electrospinning process, is described below by Dr. Nishath Khan of the Misra lab:

> The positive terminal of the high voltage supply is connected to tip of needle and the negative or ground terminal is connected to the conductive collector. As the intensity of the electrical field between the tip and collector is increased, the hemispherical surface of the fluid at the tip of the needle elongates to form a conical shape known as the Taylor cone. As the electric field is further increased, at a particular value the repulsive electrostatic force overcomes the surface tension and the charged jet of the fluid is ejected from the tip of the Taylor cone. The discharged jet undergoes an instability and elongation process; this allows

---

[2]Manjusri Misra lab, School of Engineering and the Department of Plant Agriculture, University of Guelph, Ontario.

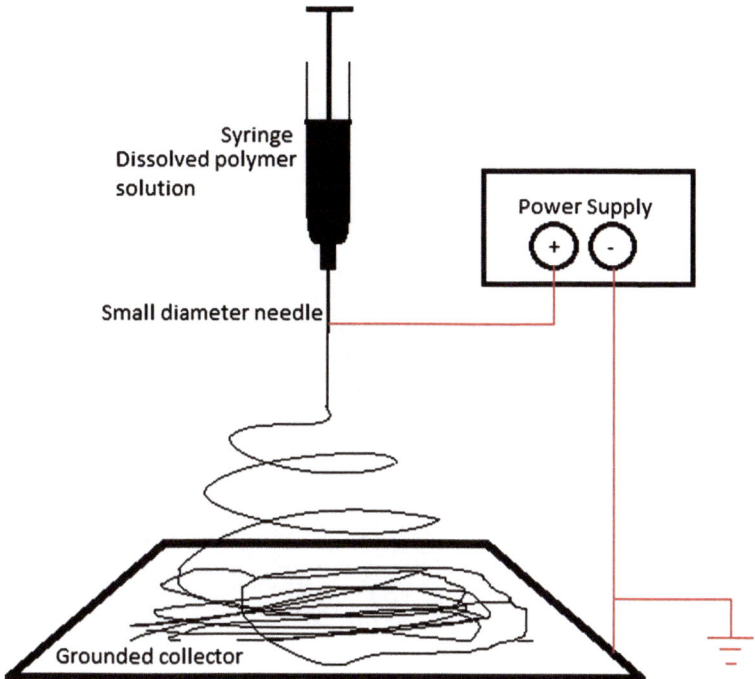

Fig. 4.3 Electrospinning overview schematic (Dr. Nishath Khan, University of Guelph)

Table 4.4 Important processing parameters for electrospinning

	Notes
Solution properties	
Viscosity (determined by concentration)	
Surface tension	
Vapor pressure	
Solvent conductivity	
Solution boiling point (determined by PLLA concentration)	Must be high enough to prevent premature solidification (which results in failure to form Taylor cone)
Setup properties	
Voltage	
Distance between tip and collector	
Type of collector	Determines the electric field between the syringe needle tip and the ground collector
Feed rate	
Type of syringe	

Source: Misra et al. (2011)

Fig. 4.4 Scanning electron microscope (SEM) image of PLA nanofiber mat with magnifications of 10 k (produced in the Misra laboratory)

the jet to become very long and reduces the diameter. The solvent in the jet evaporates and becomes dry before reaching the collector due to the high voltage and environmental conditions. Hence, the changes of environmental conditions are also very important for getting good quality of nanofibers. The solvent evaporation are also depends on the distance between the tip and collector, the solution vapor pressure and the temperature inside the chamber (Misra et al. 2011).

Several samples of 6 and 7–8% (by weight) composite fibers were produced using this technique. The final nonwoven mesh was sheet-like, with some samples being more tightly packed than others (see Fig. 4.5).

For the purpose of testing, samples of the same concentration were used interchangeably, and differences among the various iterations of the process were considered to be insignificant.

Soybean Protein Fiber

The major components of the soybean are oil, protein, and carbohydrate. Soybean protein can be isolated from the other components via extraction from de-hulled and defatted soybean flakes using alcohol or petroleum[3] (Horan 1974; Belter et al. 1944). Soybean protein can be further processed into fiber through different methods.

---

[3]Belter et al. demonstrated as early as the 1940s that extraction with alcohol was a superior method of soy protein production. Horan (1974) discussed the hexane extraction method.

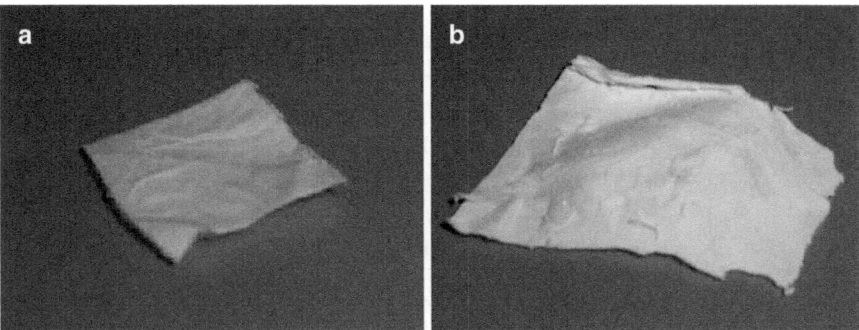

Fig. 4.5 PLLA fibers. Images of (**a**) 6% and (**b**) 7–8% PLLA nanofiber composites

Soybean protein has been used to create bone filler that rivals its commercial counterpart. The fibers are successful as both a scaffold for osteoblast growth and a catalyst of osteoblast differentiation. Soybean can also inhibit monocytes and macrophages (Giavaresi et al. 2010).

The soybean protein fiber used in this study is azlon fiber, a textile fiber made from soybean curd that has been recycled from the production of soy food products such as soymilk. The fibers are combined with cotton and bamboo-derived rayon to provide the necessary structure. This type of soy is commonly used in clothing products. All soybean materials were obtained from BabySoy, a manufacturer and retailer of renewable children's clothing. The steps involved in production of the soybean fiber are as follows (BabySoy 2007):

1. Proteins are extracted from the leftover materials of tofu or soymilk production.
2. Protein liquids are forced through a device resembling a showerhead, called a spinneret, to make liquid soy. This is called wet spinning.
3. Liquid soy is solidified to make soybean protein fiber. After protein is extracted from the leftover soybean pulp, it can be used as fertilizer.

Unlike the PLLA nanofiber composite, the soybean fiber is woven (see Fig. 4.6). Pending confirmation of biocompatibility, the extremely high porosity of this material at the nanoscale level might have implications for cell colonization on and through the material, and it may make soybean fiber an interesting candidate for consideration as a scaffold material in tissue engineering.

Fig. 4.6 36% Soybean
protein fiber

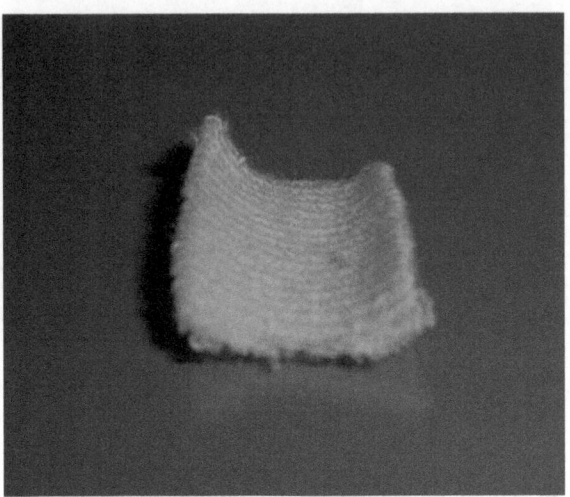

The preliminary test of biocompatibility involves incubation of a number of cell lines with the PLLA and soybean fiber meshes under controlled conditions. The relevant experimental procedures will be described in the following chapter.

Chapter 5
Feasibility Study of Corn- and Soy-Derived Materials

The first line of questioning related to the plausibility of an industry for syntheto-natural biomaterials, particularly for use in a clinical setting, is whether such materials are compatible with various human tissues. The answer to this question determines both whether further inquiry is warranted and which, if any, clinical applications are possible. Four cell lines were considered in this study (see Table 5.1).

Biocompatibility Testing Method

The biocompatibility testing protocol used in this study is as follows:

1. Obtain a 1 cm^2 square sample of material to be tested
2. Place samples in separate wells of a 6-well plate
3. Sterilize the 6-well plate under UV light for 30 minutes
4. Seed cells into each well at a density of 50,000 cells per well
5. Image cells after 24, 48, and 72 h

All cell lines except for the cardiac fibroblast line were obtained from American Type Culture Collection (ATCC). The cardiac fibroblasts were taken directly from rat hearts. Each cell line was cultured at 37 °C in a 5% (v/v) CO_2 incubator in media according to the recipes listed in Table 5.1. After growing cell lines up to a reasonable scale, cells from three of the four lines under study were incubated with the PLLA and soybean fibers in 6-well plates (see Fig. 5.1). The cardiac fibroblasts were incubated with soybean fibers.

The corn biocompatibility tests were performed using a control and two different concentrations of PLLA (6 and 7–8%). The soy tests were performed using a 36% soybean fiber material (Fig. 5.2).[1]

---

[1] 36% Soybean protein fiber, 36% cotton, 28% nylon.

O.A. Fatunde and S.K. Bhatia, *Medical Devices and Biomaterials for the Developing World: Case Studies in Ghana and Nicaragua*, SpringerBriefs in Public Health, DOI 10.1007/978-1-4614-4759-7_5, © Springer Science+Business Media New York 2012

Table 5.1 Cell lines tested and optimal growth conditions

Cell line	Culture medium[a]
Neuro-2a neuroblastoma (ATCC CCL-131)	EMEM + 10% FBS, 1% L-glutamine, 1% P/S
BCE C/D-1b Corneal endothelium (ATCC CRL-2048)	DMEM + 10% FBS, 1% P/S
PC 12-Pheochromocytoma (ATCC CRL-1721.1)	F-12K + 15% DHS, 2.5% FBS, 1% L-glutamine, 1% P/S
Cardiac fibroblast (primary harvest)	M199 + 10% FBS, MEM NEAA, Hepes, glucose, L-glutamine, penicillin, VitB12

Source: ATCC

[a]*EMEM*-Eagle's minimum essential medium, *DMEM*-Dulbecco's modified Eagle's medium, *FBS*-fetal bovine serum, *F-12K-ATCC* F12K medium, *P/S*-Penicillin/Streptomycin

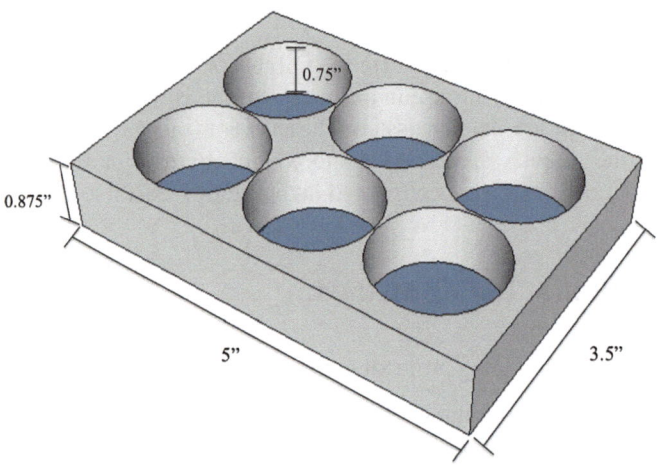

Fig. 5.1 6-well plates used for biocompatibility testing. Plate dimensions are included

Images of the cells were taken at regular intervals (see protocol). In order to obtain a more complete view of cell progress, two fields of view were captured in each well. The field locations varied depending on the location of the biomaterial fibers (Fig. 5.3).

The purpose of the test was to determine whether the viability of cells is adversely affected by the presence of a corn- or soy-based protein mesh. Images were compiled for thorough analysis. Bhatia and Yetter (2008) demonstrated that the results of visual scoring (as measured by a five-point rating scale) of biomaterial cytotoxicity are highly correlated ($r = 0.90$) with a quantitative in vitro cell viability assay[2] and

---

[2] Results were based on visual and quantitative scoring of 33 biomaterial samples each of which was tested using the 3T3 fibroblast cell line. Visual scores were based on evaluation of morphological changes such as "cell lysis, rounding, spreading, and proliferation" (Bhatia and Yetter 2008).

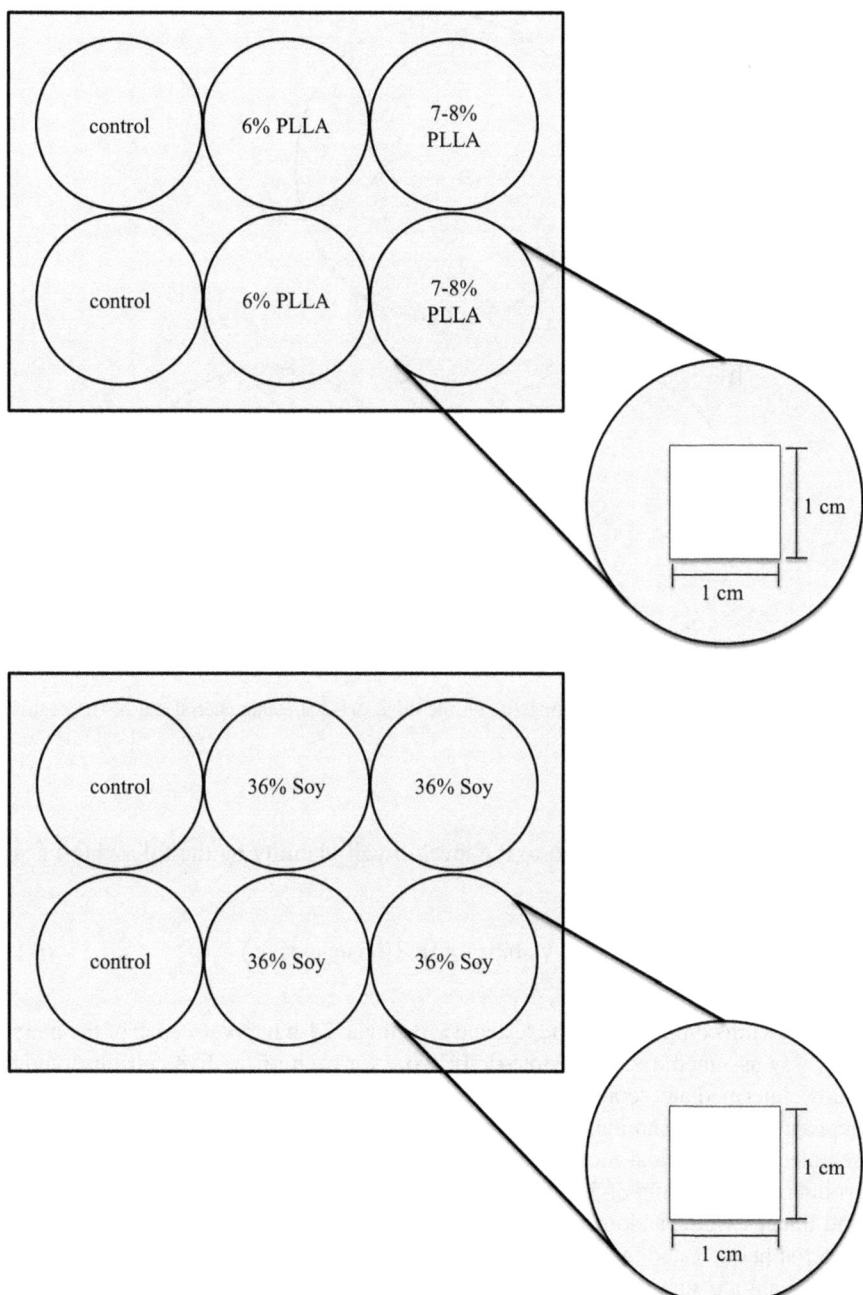

Fig. 5.2 Layout of 6-well plate used for (**a**) PLLA and (**b**) soybean biocompatibility testing. PLLA and soy pieces were initially placed in the center of the well

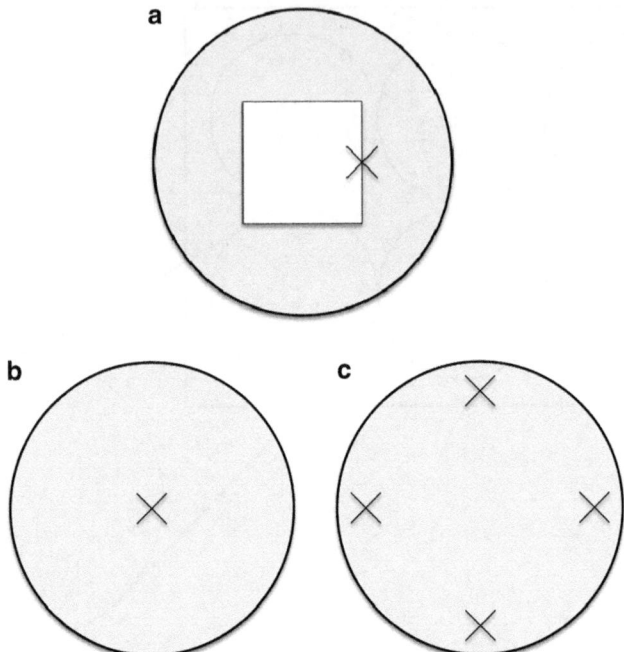

Fig. 5.3 Fields of view for imaging (**a**) on the edge of square biomaterial mesh, (**b**) central, (**c**) peripheral

that the visual score is related to the level of cell viability by the following linear equation:

$$\% \text{ Cell viability} = 18.8\,(\text{Visual score}).\tag{5.1}$$

Following completion of the test and imaging at 24-h intervals, each of the materials was assigned a visual cytotoxicity score for each of the five cell lines under study. Intermediate scores (at half-point intervals) were assigned to cells when appropriate. For example, if two samples were equally confluent but one of the samples had abnormal morphology (all other things being equal), the first sample would receive a rating of 5, and the other would receive a rating of 4.5. The samples and images were randomized, and the scorer was blinded to the identities of the material being tested in each sample. The interpretations of each rating on the five-point scale are summarized in Table 5.2. An average score was obtained for each test, and the trends in viability over the course of the test were also considered. Both the central and the peripheral fields were used to assign visual scores.

Table 5.2 Five-point rating scale for visual cytotoxicity scoring

Rating	Label	Interpretation
1	Severely cytotoxic *No visible cells*	Destruction and lysis of cells is nearly complete; considerable open areas between cells indicate that extensive cell lysis has occurred, indicating a cytotoxic reaction.
2	Significantly cytotoxic *Few visible cells*	The majority of cells are affected, but not more than 70% of the cells are rounded or lysed.
3	Moderately cytotoxic *Altered cell morphology and large gaps between cells*	Cell lysis becomes more prevalent, but no more than 50% of the cells are round and devoid of intracytoplasmic granules.
4	Mildly cytotoxic *Altered cell morphology and small gaps between cells*	Occasional lysed cells are present; not more than 20% of the cells appear to be round, loosely attached, and without cytoplasmic granules.
5	Noncytotoxic *Normal cell morphology and cell density*	Confluent monolayer of well-defined cells exhibiting cell-to-cell contact; cell morphology and cell density are not altered by the presence of a biomaterial, and discrete intracytoplasmic granules are observed. No cell lysis is observed.

Source: Bhatia and Yetter (2008)

Saturation/Movement of Square Mesh

In each case, the cells were plated above a protein mesh of either PLLA or 36% soybean fiber, located in the center of the appropriate wells. At the 48 h mark, the 6% PLLA meshes were partially saturated by the medium. By 72 h, the 6% meshes were completely saturated. However, the 7–8% materials remained dry and impermeable even after 72 h. The implication of media saturation for clinical applications is an issue in need of further investigation.

The 6% PLLA, 7–8% PLLA, and soybean fiber also displayed different patterns of movement within the well, presumably due to differing composition and density. The 6% PLLA meshes remained in the center of the well throughout the course of the study. The 7–8% meshes unanimously drifted to the edge of the well by the 72-h mark, maximizing their contact with the dry surface of the well and minimizing contact with the culture medium. The implications of this increasingly hydrophobic behavior with increasing PLLA concentration is a phenomenon worth observing in a follow-up study.

Results

Images of Neuro-2a cells are displayed below. In the interest of simplicity, pictures from the central view of a single trial are shown here.

Images

See Tables 5.3 and 5.4.

Cytotoxicity Scoring

An example of the procedure used to determine a single cytotoxicity score based on two tests per biomaterial/cell line combination is shown in Table 5.5 and Figs. 5.4 and 5.5.

Table 5.3 Neuro-2a sample images (PLLA) at 24, 48, and 72 h

Neuro-2a (h)	Control (10×)	6% PLLA (10×)	7–8% PLLA (10×)
24			
48			
72			

Dark shadows indicate that the image was taken at the edge of a PLLA or soy mesh
The scale indicated on the 24 h control picture is the same for all subsequent pictures

Table 5.4 Neuro-2a sample images (soybean fiber) at 24, 48, and 72 h

Neuro-2a (h)	Control (10×)	36% Soybean fiber (10×)
24		
48		
72		

Table 5.5 Cytotoxicity scores of three materials based on visual characterization of Neuro-2a

Neuro-2a (72 h)		6% PLLA	7–8% PLLA	36% Soybean fiber
Trial 1[a]	Central	4	3.5	3.25
	Peripheral	3	3	2.5
Trial 2	Central	4	3	3
	Peripheral	3	3	3
Average	Central	4	3.25	3.125
	Peripheral	3	3	2.75

[a]Indicates that this is the trial pictured

Images and cytotoxicity sub-scores for the remaining cell lines can be found at the end of this chapter. A summary of the average visual cytotoxicity scores for each of the three materials, as well as the corresponding cell viability level, is displayed numerically in Table 5.6. See Appendix C for a full listing of all visual scores.

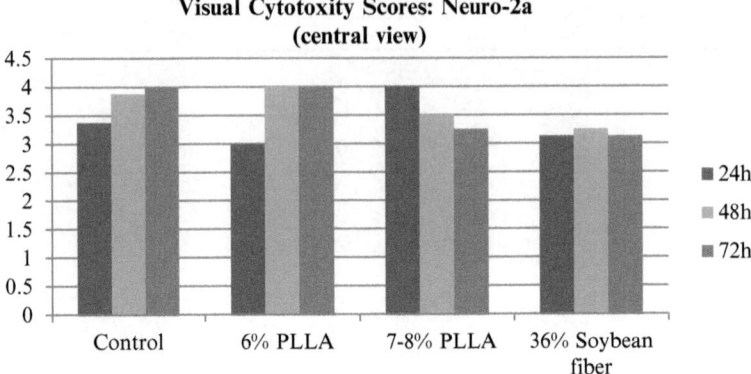

Fig. 5.4 Neuro-2a visual ratings (central)

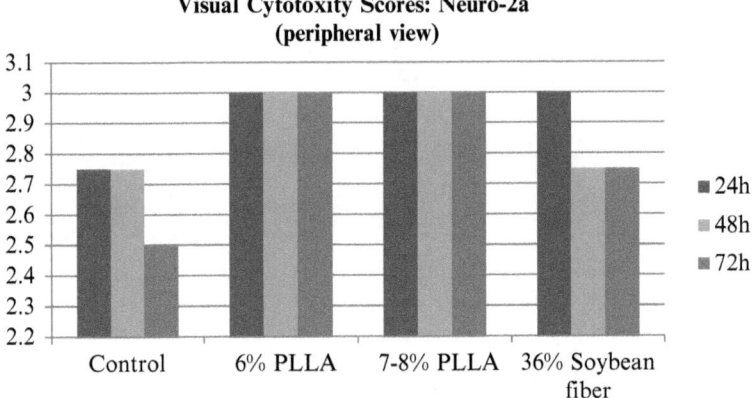

Fig. 5.5 Neuro-2a visual ratings (peripheral)

Table 5.6 Summary of cytotoxicity scores based on visual evaluation of cell morphology

Cell line		Control	6% PLLA	7–8% PLLA	36% Soybean fiber
			Type of biomaterial		
Neuro-2a neuroblastoma	Score	4	4	3.25	3.125
(ATCC CCL-131)	Viability (%)	75.2	75.2	61.1	58.75
Corneal endothelium	Score	2.75	3.25	3	3
(ATCC CRL-2048)	Viability (%)	51.7	61.1	56.4	56.4
PC 12-Pheochromocytoma	Score	5	3.5	4.25	3.75
(ATCC CRL-1721.1)	Viability (%)	94	65.8	79.9	70.5
Cardiac fibroblast	Score	3.75	–	–	3.5
(primary harvest)	Viability (%)	70.5	–	–	65.8

Discussion

The Neuro-2a test pictured above results in final average visual toxicity scores in the center of the well (after 72 h) of 4, 3.25, and 3.125 for a 6% PLLA mesh, a 7–8% PLA mesh, and a 36% woven soybean protein fiber, respectively. At first glance, these scores raise concern about the proliferation of cells in the presence of the biomaterials. However, in each case the average was within a single point of the average score for the control well, suggesting that the biomaterial fibers, independent of the culture conditions, had little effect on cell growth and spreading. Peripheral cytotoxicity ratings as shown in Fig. 5.5 suggest that both concentrations of PLLA meshes are noncyctotoxic to the Neuroblastoma tissues (relative to the control) and that this is maintained over the 72 h of observation. Several additional tests must be performed to establish complete biocompatibility of corn-based meshes with neural cells. The soy mesh initially displays results similar to those of the PLLA, but after 24 h its cytotoxicity increases, and its score stays relatively constant until the end of the 72-h observation period. This is a similar pattern to the control; however, the average score at each point in time is higher than the control, suggesting that the soybean fiber does not lower cell viability.

Similarly, the other three cell lines (images and scores at the end of this chapter) demonstrate that the presence of the test materials has limited influence on cell viability. For the BCE C/D-1b corneal endothelium line, the visual scores for the three biomaterials are all within one point of the control (Fig. 5.6). In fact, the average scores in the center of the well for each material after 72 h are all slightly higher than that of the control well at the same point in time for the central view.

The PC-12 cell line, on the other hand, shows more variability, with almost two points separating the visual score for the control well and the lowest biomaterial score (6% PLLA). However, this line has the highest average visual scores, indicating that even those samples that deviated significantly from the control experienced low levels of cytotoxicity.

Finally, the cardiac fibroblasts did not display significantly decreased viability in the presence of soybean fibers. These results suggest that corn PLLA and soybean fibers have insignificant effects on cell viability for the four cell lines under study. In other words, the biomaterials tested did not have an adverse effect on cell growth and proliferation. This positive result bodes well for further inquiries into clinical adaption of naturally derived biomaterials.

Additional Considerations

This study employed the visual evaluation techniques developed by Bhatia and Yetter (2008), which demonstrated that a strong correlation exists between visual *in vitro* cytotoxicity ratings and quantitative *in vitro* cell viability measurements. However, their method (and the linear relationship that resulted from it) was developed using the NIH/3T3 embryonic fibroblast line, which is excluded from this study. A more thorough feasibility study would first seek to find a comparable linear relationship for each of the cell lines under study.

Additionally, many of the scores obtained during the feasibility study were mid-range scores (between 2.5 and 4). Bhatia and Yetter advised a cautious approach to interpretation of scores near the center of the scale. In the present study, a larger number of samples are needed to gain confidence in the correlation between the visual scores and a quantitative cell viability score. Including more samples would also allow us to perform more useful statistical evaluations with the data obtained.

Finally, the majority of the cell lines studied displayed remarkable consistency. For a given cell line, most samples received a visual rating within one point of the rating for the control. In other words, the ratings with different materials lay within a tight band, although the absolute level of the band was often lower than desired. This suggests that any decrease in visual scores may have been a result of culture conditions rather than the adverse effect of the biomaterials. Further optimization of culture conditions is therefore expected to result in higher scores for both the control and test samples.

Detailed Results

See Figs. 5.6–5.11 and Tables 5.7–5.14.

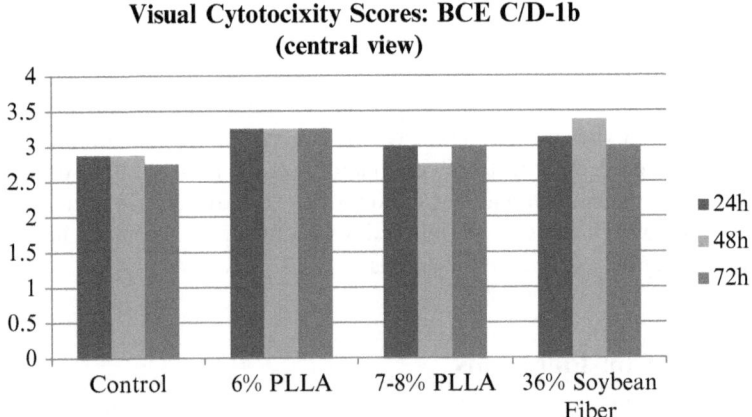

Fig. 5.6 Corneal endothelium visual ratings (central)

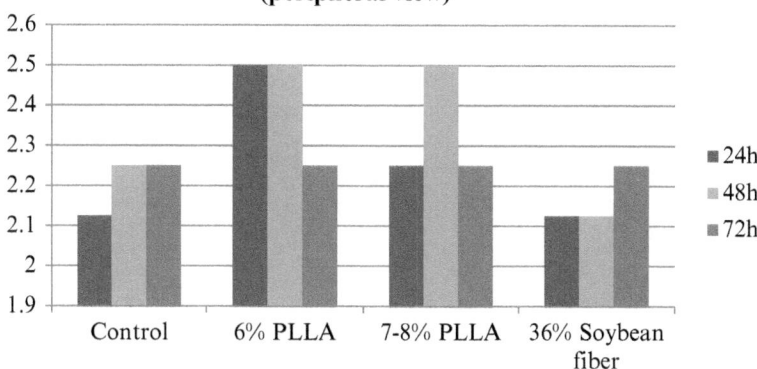

Fig. 5.7 Corneal endothelium visual ratings (peripheral)

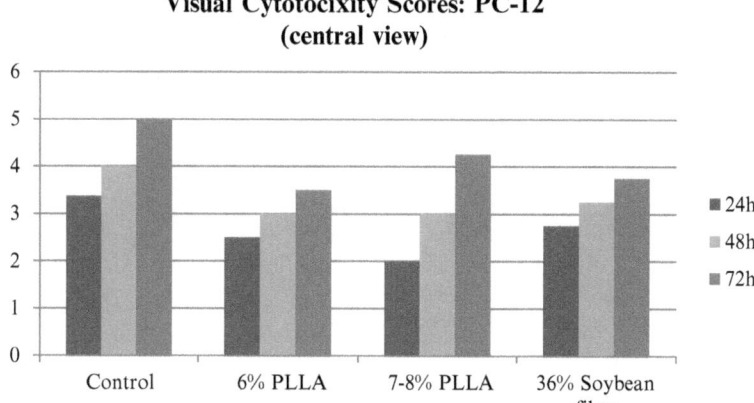

Fig. 5.8 PC-12 visual ratings (central)

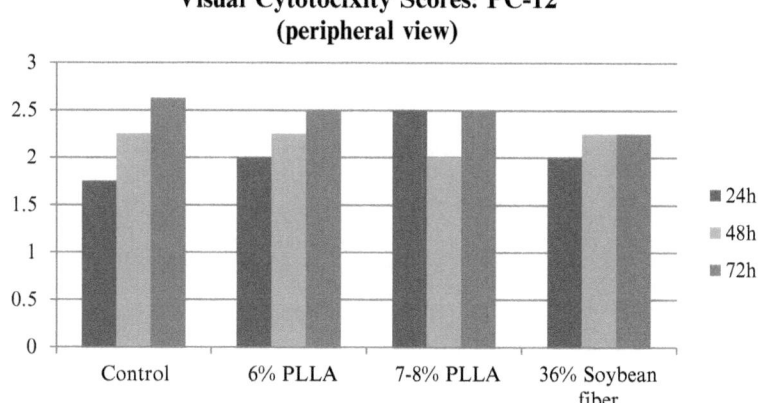

Fig. 5.9 PC-12 visual ratings (peripheral)

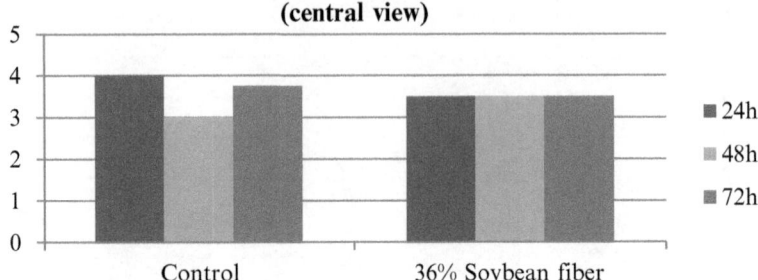

Fig. 5.10 Cardiac fibroblast visual ratings (central)

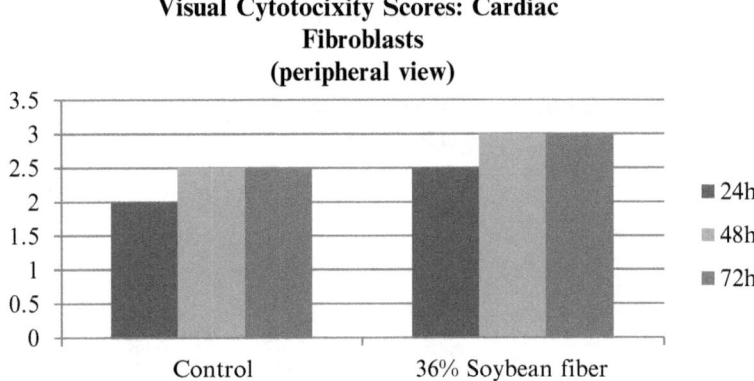

Fig. 5.11 Cardiac fibroblast visual ratings (peripheral)

Table 5.7 Corneal endothelium sample images (PLLA) at 24, 48, and 72 h

BCE C/D-1b (h)	Control (10×)	6% PLLA (10×)	7–8% PLLA (10×)
24			
48			
72			

Table 5.8 Corneal endothelium sample images (PLLA) at 24, 48, and 72 h

BCE C/D-1b (h)	Control (10×)	36% Soybean fiber (10×)
24		
48		
72		

Table 5.9 Cytotoxicity scores of three materials based on visual characterization of corneal endothelium cells

BCE C/D-1b (72 h)		6% PLLA	7–8% PLLA	36% Soybean fiber
Trial 1*	Central	4	3	3
	Peripheral	2	2.5	2.5
Trial 2	Central	2.5	3	3
	Peripheral	2.5	2	2
Average	Central	3.25	3	3
	Peripheral	2.25	2.25	2.25

*Indicates that this is the trial pictured

Table 5.10 PC-12 sample images (PLLA) at 24, 48, and 72 h

PC-12 (h)	Control (10×)	6% PLLA (10×)	7–8% PLLA (10×)
24			
48			

Table 5.10 (continued)

PC-12 (h)	Control (10×)	6% PLLA (10×)	7–8% PLLA (10×)
72			

Table 5.11 PC-12 sample images (soybean fiber) at 24, 48, and 72 h

PC-12 (h)	Control	36% Soybean fiber
24		
48		
72		

Table 5.12 Cytotoxicity scores of three materials based on visual characterization of PC-12 cells

PC-12 (72 h)		6% PLLA	7–8% PLLA	36% Soybean fiber
Trial 1	Central	3	4.5	4.5
	Peripheral	2.5	2.5	2.5
Trial 2*	Central	4	4	3
	Peripheral	2.5	2.5	2
Average	Central	3.5	4.25	3.75
	Peripheral	2.5	2.5	2.25

*Indicates that this is the trial pictured

Table 5.13 Cardiac fibroblast sample images (soybean fiber) at 24, 48, and 72 h

Primary harvest cardiac fibroblasts (h)	Control	36% Soybean fiber
24		
48		
72		

Table 5.14 Cytotoxicity scores of three materials based on visual characterization of primary harvest cardiac fibroblasts

Cardiac fibroblasts (72 h)		Control	36% Soybean fiber
Trial 1*	Central	4	–
	Peripheral	3	–
Trial 2	Central	3.5	3.5
	Peripheral	2	3
Average	Central	3.75	3.5
	Peripheral	2.5	3

*Indicates that this is the trial pictured

Chapter 6
Discussion, Recommendations, and Conclusion

The preceding five chapters have offered an in-depth discussion of the major challenges facing the successful implementation of medical devices in the context of two developing nations. Included is a preliminary investigation into a possible source of novel materials that are both sustainable and naturally derived. The questions that originally drove this exploration can now be considered specifically for the cases of Ghana and Nicaragua and in the context of the laboratory study performed in Chap. 5.

1. *What are the major factors limiting the effectiveness of medical technology in developing settings?*
2. *How can the local resources of these settings be harnessed to reverse this trend?*

There was an initial baseline observation that medical devices do not adequately meet the needs of health facilities, clinicians, and patients in the developing world. Despite a renewed focus on primary health care in recent decades, disparities in quality of care and mortality rates due to certain diseases persist. The sight of rooms filled with abandoned and nonfunctioning equipment is not an uncommon sight in developing world hospitals.[1] From these and other reports of discontinued equipment use, an image emerges of developing world clinics as black boxes where perfectly good equipment fails due to poor conditions.

Among the long-standing hypotheses for this consistent failure are the lack of training and local unavailability of spare parts where needed. In other words, traditional theories have focused on "lack" of the appropriate supporting materials to make equipment function as intended. However, a closer look reveals that the problem lies less at the final destination than in the actual process and the criteria used to determine which equipment should be donated by whom to which locations. The geographical (including climate) and economic conditions of the developing world are such that the qualities that make technology and instrumentation successful in

---

[1] See Appendix A.

O.A. Fatunde and S.K. Bhatia, *Medical Devices and Biomaterials for the Developing World:*   83
Case Studies in Ghana and Nicaragua, SpringerBriefs in Public Health,
DOI 10.1007/978-1-4614-4759-7_6, © Springer Science+Business Media New York 2012

low-resource environments are inherently quite different, and in some sense opposite, from those that are celebrated in the industrialized world. The maximally digitized, automated nature of today's technological advances simply has no place under certain conditions. In the absence of economic incentives for device manufacturers to prioritize design principles that are appropriate for the developing world, it is the role of governments and researchers to determine how the technical capabilities of the developing world can be used to address such issues from the inside out.

Medical Technology Score

The search for an appropriate solution requires quantification of the problem, or expression in a common language of how well or poorly a certain nation's level of health technology compares to the ideal given its demographic, geographic, and economic situation. Our proposed solution to this need is the adoption of a standard metric called the Medical Technology Score (MTS).[2] The MTS, which is designed to be computed at the facility level, aims to capture certain important characteristics, such as adherence to a standard inventory and preparation to treat locally relevant conditions. The power in such a metric lies in the fact that it empowers international policymakers to make internal and cross-country comparisons of interest, which can lead to development of appropriate interventions based on patterns within related groups of countries.

The major drawback to the creation of this metric is its reliance on a large amount of reliable data, which is not uniformly available. Because of the resources required to collect this information and the importance of standardized collection for comparability of the metric among countries, the scale and organization of an international body such as the WHO would be beneficial for the undertaking of such a project.

The MTS also aims to quantify "level of technology" strictly based on inventory. However, there are a number of unquantifiable factors, particularly those related to distribution of facilities and of the population, which surely impact the relative need for equipment in different areas. An ideal metric would control for this and other qualities. The MTS is only a preliminary attempt at the task of understanding trends in technology, and it will require significant refining.

The Role of Biomaterials

While it is essential to challenge the current one-way flow of technology into the developing world by quantifying the lack of technology in specific nations, a sustainable solution requires the parallel strengthening of internal capacity. Coloma and

---

[2] See Appendix B for the conceptual development of the MTS. See Chaps. 2 and 3 for examples of its use.

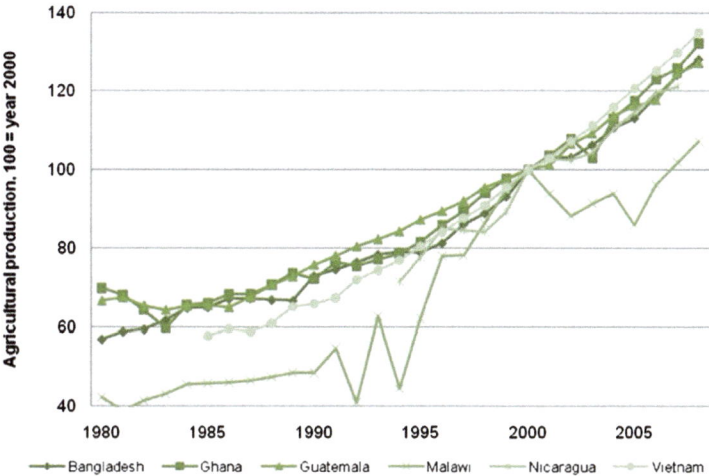

Note: Agricultural value added measured in constant 2 000 USD. Index for each country is set to 100 for year 2000.
Source: World Development Indicators.

Fig. 6.1 Agricultural production in DEVPEM countries, 1990–2008 (Brooks et al. 2011)

Harris (2008) instituted a system for this exact purpose in several Central American nations through their work with the Sustainable Sciences Institute. However, their focus was mostly on establishment of biomedical laboratory capacity. Use of such techniques with transfer of large equipment would undoubtedly pose logistical and financial difficulties that would threaten the long-term sustainability of the model. In the decades to come, it must become a topmost priority of would-be technology pioneers to engage local engineering institutions in a manner that spurs innovation in accordance with local needs and resources.

Our explanation for the current unidirectional flow is that technology has traditionally been equated to electronics, which largely originate in the West and other advanced societies. If we broaden the definition of technology, we see that countries in the developing world have unique potential in the development of novel technology because of their agricultural output and the burgeoning role of agriculture in transforming their economies (Fig. 6.1).

The identified "need" for biomaterials in the developing world is based on the potential for these biomaterials to improve quality of care for health problems that represent a huge burden of disease in developing countries. Scientists from a range of specialties must be encouraged to conduct further research into novel uses of syntheto-natural biomaterials, which will hopefully provide additional impetus for strengthening of health systems and both research and manufacturing infrastructure. Elling indicated that three critical planning aspects of an ideal health system include "a comprehensive health plan, coordination with national goals other than health,

Table 6.1 Methods for characterization of biomaterials

Mechanical characterization	*Biological characterization*
Mechanical strength	Sterility properties
Tensile strength	Bioburden
Shear strength	Bacterial endotoxin assay
Impact strength	Tissue compatibility
Cohesive and adhesive strength	Cytotoxicity
	Cellular inflammation
Physical/chemical characterization	Cell and protein attachment
Curing and reaction properties	Tissue irritation
Extent of reaction	Tissue implantation response
Residual starting materials	Wound healing
Heat of reaction	Hemolysis testing
Degradation properties	Systemic effects
Degradation rate	Pyrogenicity
Degradation products	Sensitization
Swelling determination	Toxicokinetic evaluation
Drug release properties	Metabolic fate
Drug delivery rate	Antimicrobial effects
Drug bioactivity	Encapsulated live cell viability
Device characterization	*Clinical characterization*
Accelerated shelf stability test	Ease of use
Physical integrity	Patient and clinician acceptance
Device functionality	Clinical efficacy
Device preparation time	Cost-effectiveness

Source: Bhatia (2010)

and broad popular participation" (Donahue 1991). Neither Ghana nor Nicaragua has a national health technology policy (WHO 2010c). In order for this idea to come to fruition, a necessary next step is the formalization of bodies devoted to this topic. A number of additional considerations such as financing and scalability of biomaterial usage would be the responsibility of such a body. It is also important to consider the cultural implications of converting materials that have traditionally served as food products into medical devices. Finally, formal collaboration with non-health sectors in target countries will be required.

We have preliminary evidence of biocompatibility between implantable corn- and soy-derived materials and selected cell lines, but the study completed in Chap. 5 was only the first of a long list of evaluations that must be undertaken. Bhatia (2010) provides a comprehensive list of the parameters along which biomaterials must be characterized before a decision can be reached about their usefulness and potential impact (see Table 6.1).

The large number of investigations that are still required raises questions about the level of infrastructure that must be put into place before these materials are even approved for medical use. It is also important to remember that the scientific considerations involved are only a subset of the requirements for implementation.

The feasibility studies included in this text are thus encouraging, but by no means conclusive or predictive of reality. It is at this juncture that collaboration between government, the private sector, and (local and remote) research institutions is required in order to ensure that the appropriate steps are taken towards development, production, and regulation of these materials.

Medical Applications

The rapid advance of technology is often such that new developments build upon themselves, leaving behind the original intent and desired applications of science. As a discipline, engineering, which aims to be "science applied to the solution of human problems," must strive to bear in mind potential applications of novel discoveries throughout the development process. This underlying goal of advancing human health is somewhat implicit in the investigation of biomaterials for clinical applications. An important concern of countries being presented with such a proposal should logically be the possible applications of biomaterials in local clinical settings. A general discussion of biomaterial applications is completed in Chap. 4, but a more targeted approach should take into account the needs of individual nations.

Ghana and Nicaragua have the following conditions common to their "top-ten" list: stroke, coronary heart disease, influenza and pneumonia, lung disease, road traffic accidents, and kidney disease (Table 6.2). A close look at the remaining unique diseases suggests that the burden of disease in Ghana remains heavily impacted by infectious diseases, while chronic and noncommunicable diseases have assumed the forefront of health concerns in Nicaragua.

Of these six conditions, five have pre-identified needs that can be met by biomaterials that are currently under study or already in use (Bhatia 2010), and four are included in the WHO Availability Matrix (Hansen et al. 2010). This suggests that

Table 6.2 Lists of "high mortality" diseases

Cause (Ghana)	Rank	Cause (Nicaragua)	Rank
Diarrheal diseases	1	**Coronary heart disease**	1
Stroke	2	Diabetes mellitus	2
Coronary heart disease	3	**Stroke**	3
HIV/AIDS	4	**Kidney disease**	4
Influenza and pneumonia	5	Liver disease	5
Tuberculosis	6	**Influenza and pneumonia**	6
Lung disease	7	**Lung disease**	7
Malaria	8	Hypertension	8
Road traffic accidents	9	**Road traffic accidents**	9
Kidney disease	10	Stomach cancer	10

Bolded conditions are common to Ghana and Nicaragua

international bodies have recognized the importance of equipping facilities to treat diseases that impact the developing world. This strategy is consistent with the goal of integrating care of infectious diseases into primary health care, particularly in areas where specialists are lacking. Evaluation of inventories (using a metric such as the MTS) and testing of biomaterials for clinical use should use this concept as a guide by prioritizing treatment of conditions that contribute significantly to mortality in specific regions.

Agriculture

The feasibility of corn- and soy-derived materials, each of which is developed from a product of a naturally occurring organism, was the subject of Chap. 5. Maize is among the largest crops (in terms of production) in Ghana. While the soybean is not produced on a commercial scale, Ghana has access to this crop via regional trade within West Africa. Nicaragua's internal access is more limited because of a need for nearly complete conversion of crop products into food, but the nation has access to additional quantities (much more than Ghana) via trade within Mesoamerica and with the United States.

The availability of these materials via agricultural production, similar in many ways to the *in vitro* biocompatibility of their derivatives with clinically relevant cells lines, is only the gatekeeper to a number of criteria that must be met before the necessary agriculture-to-health conversion becomes viable (Fig. 6.2). The extraction of PLLA from starch, processing of PLLA via electrospinning, and manufacture of implantable devices using PLLA are all capabilities that would need to be present in order for the corn-derived materials discussed here to become usable for clinical applications. An additional concern is whether the clinical procedures that would make use of these materials occur frequently enough to make the in-country manufacture of such devices economically feasible (excluding the possibility of production for export, which comes with its own challenges).

A majority of the agriculturally derived household income in Ghana comes from crop production. On the other hand, agricultural household income in Nicaragua is more diverse, resulting from a mixture of crops, livestock, and wage employment (Table 3.5). This reflects the slight distinctions between the economies of the two nations. Furthermore, it suggests that a crop-targeted strategy for linking the agricultural and health sectors might have more impact in a country like Ghana. In both the public and private sectors, Ghana has a lower median availability of essential medicines (WHO 2012), making it an interesting candidate for a pilot version of the strategy described above. Consideration of such factors must be included in a strategy for the industrial manufacture of biomaterials, which should be coordinated among all of the affected sectors in each country.

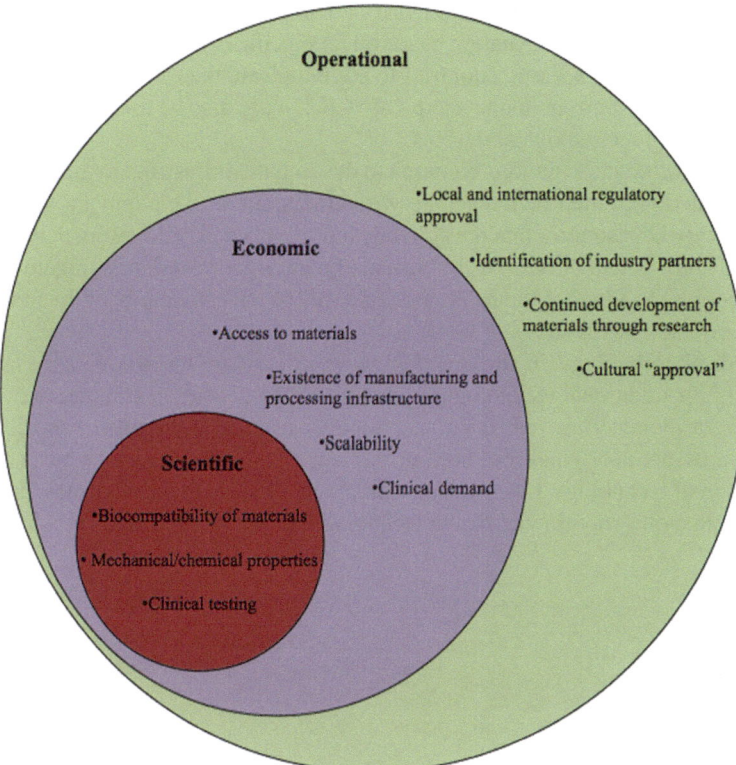

Fig. 6.2 Levels of feasibility of biomaterials

Concluding Remarks

The topics presented in this book leave much to be considered and an equal amount to be pursued through further research. Our argument is fourfold:

First, it has become clear that medical equipment (in general) is ineffective in the developing world, limiting the maximum possible quality of care available. Part of the reason for the continued limitation is the dependence of the developing world on equipment developed for and originating from the developing world, where the criteria for appropriateness are inherently different.

Secondly, we challenge the idea, which is supported by this one-way flow, that there is no potential for the developing world to contribute to the advancement of technology. This is where the idea of changing the prevailing definition of technology comes into play. A challenge that must be overcome in order for developing nations to realize their possible contribution is the improvement of health care capacity within the context of their overall development programs.

Thirdly, we propose the development of the MTS as a metric that summarizes the technological standing of a nation based on its specific needs using different parameters. This will help specific countries to communicate their needs more clearly. In the context of the current donor–recipient model, it can also be used to match donors to recipients of technological devices.

Finally, while these nations continue to develop the infrastructure needed to support production of their own machinery/traditional medical equipment, they should also look to the resources that they already have in place. By doing so, they uncover their inherent potential (such as in the use of naturally derived, or syntheto-natural, biomaterials) to advance the frontier of science, which can help both the industrialized world and the developing world.

Medical technology in the developing world can be transformed by building capacity for traditional technologies via engineering education and research and by building a biomaterials industry that reroutes agricultural output for medical use. This represents an original contribution to science if we operate under a much broader definition of technology. Over time, development of such an industry may transform the oft-cited one-way flow of health technology into a two-way dialogue.

Appendix A
Images from Health Facilities in Nicaragua

Each of the devices pictured on the following two pages is nonfunctioning and housed either in its original location or in an abandoned "equipment warehouse." Pictures were taken at health facilities in the cities of Managua and Ocotal.

(a) Refrigerator
(b) Cold box (for vaccinations)
(c) Drill
(d) Spectrophotometer
(e) Syringe infusion pump
(f) Misaligned X-ray machine
(g) Blood sample analyzer
(h) Assorted disassembled electronics
(i) Assorted parts
(j) Assorted parts
(k) Infant incubator
(l) Anesthesia machine
(m) Tissue culture incubator
(n) Assorted parts

O.A. Fatunde and S.K. Bhatia, *Medical Devices and Biomaterials for the Developing World: Case Studies in Ghana and Nicaragua*, SpringerBriefs in Public Health, DOI 10.1007/978-1-4614-4759-7, © Springer Science+Business Media New York 2012

Appendix B
Medical Technology Score

Qualities That the MTS Aims to Account for

(For a single facility)

- Whether a facility has the right equipment (right is defined based on the facility type)
- Whether a facility has the right number of each device (with emphasis on the most important devices)
- Whether a facility is equipped to treat conditions that are locally relevant

(Other goals)

- Ease of comparison between nations, taking into account the unique distribution of facilities in each

Possible Approaches

1. The facility score is the average of individual item scores, weighted by the level of importance of each item. The category score is the average of facility scores, weighted by the "weight" of each category in that country.

   (a) This puts the scores all over the same denominator so that they can become percentages, making interpretation easier.

2. The score is an aggregate of "points." Facilities receive extra points for desirable qualities (one point each for items A and B, two for C because it's more important).

   (a) There is a problem here—it's boundless! It could be huge for bigger hospitals because they have more equipment (but not necessarily a larger proportion of what they should have).

O.A. Fatunde and S.K. Bhatia, *Medical Devices and Biomaterials for the Developing World:*   95
Case Studies in Ghana and Nicaragua, SpringerBriefs in Public Health,
DOI 10.1007/978-1-4614-4759-7, © Springer Science+Business Media New York 2012

Challenges

1. Not every country has a detailed and up-to-date comprehensive inventory, and those that do collect data in different ways.
2. Does this totally account for regional differences/population distribution? (Are these already implicit in the distribution of health facilities, which are to some extent defined on an urban–rural continuum?)

How Do We Define the Following?

1. Equipment? What do we call them?

   – *Equipment are identified using unique five-digit codes from the Global Medical Device Nomenclature.*

2. "Importance" of one device as compared to another?
   – *This will be captured using criteria of inclusion of medical devices, which prioritizes them according to function, rate of use, risk involved, and maintenance. There are three different models (WHO 2010a):*

   Fennigkoh and Smith model (equipment management number):
   1. EMN = Function + Risk + Required maintenance

   Wang and Levenson's Equipment Management Rating:
   2. EMR = "Mission Critical Rating" + 2 × Risk + 2 × Maintenance
   3. Adjusted EMR = ("Mission Critical Rating" + 2 × Maintenance) × Utilization + 2 × Risk

   For simplicity, the Fennigkoh and Smith model was used in this analysis.

3. Ideal inventory for a given facility?

   – *The WHO has published preliminary guidelines, entitled "Medical Devices by Healthcare Facilities."*

4. Facility classifications?

   – *The WHO and UNICEF classify health facilities based on size and services provided. These guidelines should be used to make consistent assignments of different facilities to appropriate categories.*

5. Which diseases should be treatable in a given nation?

   – *This includes several conditions, but the focus is on the top causes of mortality for nations under study.*

6. What equipment is required to treat a given disease?

   – *The WHO Availability Matrix provides information on which equipment is required to adequately treat 15 major diseases.*

Appendix C
Calculation of the Essentiality Score (*e*)

The measure of "essentiality" for a specific piece of medical equipment, denoted by the variable *e*, is based on the model developed by Fennigkoh and Smith WHO (2110a) in the WHO's "Criteria for medical inventory inclusion."[1] This model assigns points to pieces of equipment based on their function, the risk associated with their use, and their maintenance requirements. Table C.1 details the characteristics

Table C.1 Equipment management number components

Category	Description	Point score
Function		
Therapeutic	Life support	10
	Surgical and intensive care	9
	Physical therapy and treatment	8
Diagnostic	Surgical and intensive care monitoring	7
	Additional physiological monitoring and diagnostic	6
Analytical	Analytical laboratory	5
	Laboratory accessories	4
	Computers and related	3
Miscellaneous	Patient related and other	2
Risk		
	Potential patient death	5
	Potential patient or operator injury	4
	Inappropriate therapy or misdiagnosis	3
	Equipment damage	2
	No significant identified risk	1
Maintenance		
	Extensive: routine calibration and part replacement required	5
	Above average	4
	Average: performance verification and safety testing	3
	Below average	2
	Minimal: visual inspection	1

Source: Fenningkoh and Smith (2010)

---

[1]This document is part of the WHO Medical device technical series.

O.A. Fatunde and S.K. Bhatia, *Medical Devices and Biomaterials for the Developing World: Case Studies in Ghana and Nicaragua*, SpringerBriefs in Public Health, DOI 10.1007/978-1-4614-4759-7, © Springer Science+Business Media New York 2012

that correspond to different scores in each category. While this model places twice as much weight on function as on risk and maintenance, other models (such as Wang and Levenson, WHO 2010a) offer alternatives that weight the three parameters more evenly. These models were originally developed for the purpose of determining whether or not to include a certain medical device in an inventory, but their usefulness can be generalized to evaluation of existing inventories against a standard.

Adding the scores in the three categories results in an equipment management number (EMN), which can range from 4 to 20.

$$EMN = Function + Risk + Required\ maintenance. \qquad (C.1)$$

The EMN for a given device is converted into an "essentiality score" simply by dividing its EMN by 20 (the highest possible EMN) in order to obtain a score in percentage form.

$$e = \frac{EMN}{20}. \qquad (C.2)$$

Appendix D
Visual Cytotoxicity Ratings

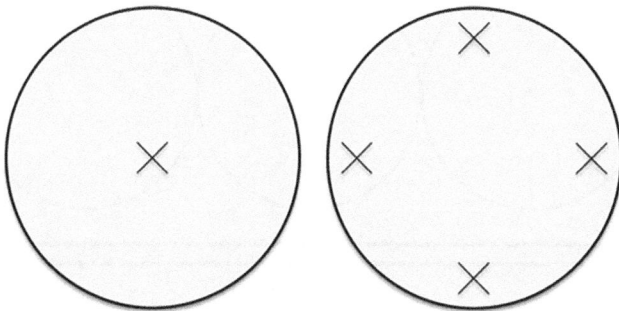

Fig. D.1 Imaging locations within 6-well plate. Each field was coded as either (**a**) central or (**b**) peripheral

Coding of samples

Each sample was assigned a four-digit code that was unique among other samples from the same cell line. The codes were generated as follows:

Four-digit code:

$$W\,X\,Y\,Z$$

where:

W designates the plate identity, X and Y together identify the well number and field of view, and Z identifies the time elapsed since incubation. More specifically,

W is equal to 1 or 2 depending on the biomaterial being incubated in a given plate ($1 = $ PLLA, $2 = $ soy).

XY is a two-digit number that identifies the field of view (see Fig. D.1). XY ranges from 01 to 12, with odd numbers corresponding to central fields and even numbers corresponding to peripheral fields (Fig. D.2).

O.A. Fatunde and S.K. Bhatia, *Medical Devices and Biomaterials for the Developing World:*   99
Case Studies in Ghana and Nicaragua, SpringerBriefs in Public Health,
DOI 10.1007/978-1-4614-4759-7, © Springer Science+Business Media New York 2012

Z is equal to 1, 2, or 3 depending on the time at which the sample was collected
(1 = 24 h, 2 = 48 h, 3 = 72 h).

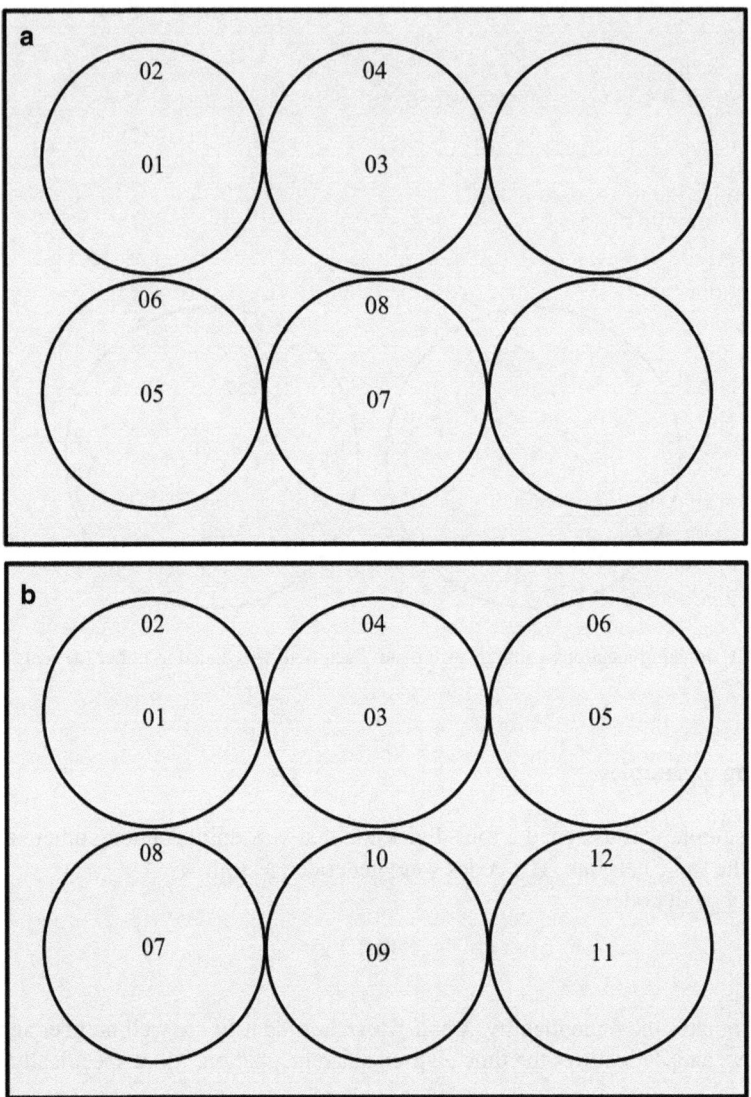

Fig. D.2 Field-of-view codes for (**a**) cardiac fibroblasts and (**b**) all other cell lines

For example, a sample co-incubated with PLLA and that was collected at 48 h
from the center of well 4 would be assigned a code of 1072.

Figure D.3 shows the plate location associated with each code. Table D.1 provides the visual scores assigned to each field, listed by code.

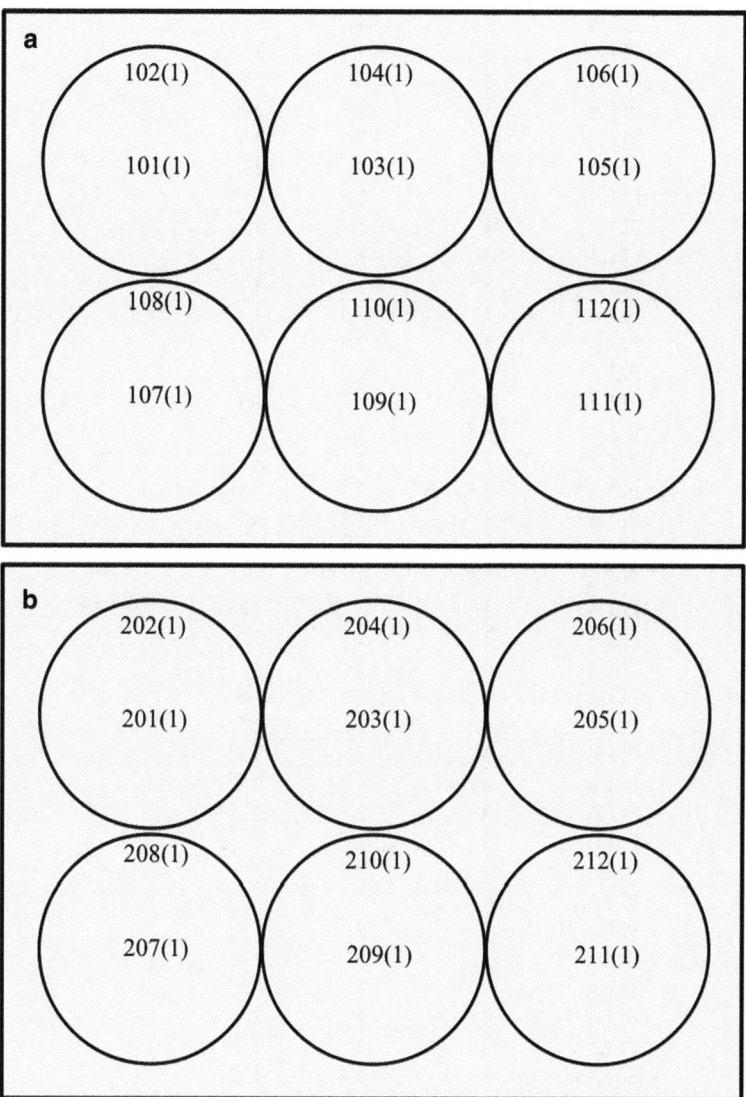

Fig. D.3 Map of sample codes for cell lines (other than cardiac fibroblasts) incubated with (**a**) PLLA and (**b**) soybean fiber. The codes pictured reflect samples that were collected at 24 h

Table D.1 Summary of visual cytotoxicity scores

Sample code	Biomaterial	Cell line			
		Neuro-2a	PC-12	BCE C/D-1b	Cardiac fibroblasts
1011	PLLA*	4	4	2	4
1012	PLLA*	4	4	4	3
1013	PLLA*	4	5	3.5	4
1021	PLLA*	3	2	2.5	2
1022	PLLA*	3	2	2.5	3
1023	PLLA*	2	3	3	3
1031	PLLA*	4	2.5	4	3.5
1032	PLLA*	4	3	3.5	–
1033	PLLA*	4	3	4	–
1041	PLLA*	3	2	3	2
1042	PLLA*	3	2.5	2	–
1043	PLLA*	3	2.5	2	–
1051	PLLA*	4	2	3	4
1052	PLLA*	4	3	2.5	3
1053	PLLA*	3.5	4.5	3	3.5
1061	PLLA*	3	3	2	2
1062	PLLA*	3	2	2.5	2
1063	PLLA*	3	2.5	2.5	2
1071	PLLA*	3	3	4	3.5
1072	PLLA*	3.5	4	3.5	3.5
1073	PLLA*	4	5	3.5	3.5
1081	PLLA*	2	1	2	3
1082	PLLA*	3	2	2	3
1083	PLLA*	3	2.5	2.5	3
1091	PLLA	2	2.5	2.5	
1092	PLLA	4	3	3	
1093	PLLA	4	4	2.5	
1101	PLLA	3	2	2	
1102	PLLA	3	2	3	
1103	PLLA	3	2.5	2.5	
1111	PLLA	4	2	3	
1112	PLLA	3	3	3	
1113	PLLA	3	4	3	
1121	PLLA	3	2	2.5	
1122	PLLA	3	2	2.5	
1123	PLLA	3	2.5	2	
2011	Soybean	3	3.5	4	
2012	Soybean	4	4	4.5	
2013	Soybean	4	5	3.5	
2021	Soybean	3	2	2	
2022	Soybean	2	2.5	2	
2023	Soybean	2	2.5	2	
2031	Soybean	3.5	3	4	
2032	Soybean	3	3	4	
2033	Soybean	3.5	4	3.5	

(continued)

Table D.1 (continued)

Sample code	Biomaterial	Cell line			
		Neuro-2a	PC-12	BCE C/D-1b	Cardiac fibroblasts
2041	Soybean	3	2	2.5	
2042	Soybean	3	2.5	2.5	
2043	Soybean	2	2.5	2.5	
2051	Soybean	3	3	3	
2052	Soybean	4	4	3	
2053	Soybean	3	5	2.5	
2061	Soybean	3	2	2	
2062	Soybean	3	2	2	
2063	Soybean	3	2.5	2.5	
2071	Soybean	3.5	3	4.5	
2072	Soybean	4	4	4	
2073	Soybean	4	5	4	
2081	Soybean	3	2	2	
2082	Soybean	3	2.5	2	
2083	Soybean	3	2.5	2	
2091	Soybean	3	2.5	3	
2092	Soybean	3	3	3.5	
2093	Soybean	3	3	3.5	
2101	Soybean	3	2	2	
2102	Soybean	2	2	2	
2103	Soybean	3	2	2	
2111	Soybean	3	2.5	2.5	
2112	Soybean	3	3	3	
2113	Soybean	3	3	2.5	
2121	Soybean	3	2	2	
2122	Soybean	3	2.5	2	
2123	Soybean	3	2	2	

The *starred rows* indicate fields where PLLA was the biomaterial used for three of the four cell lines. Cardiac fibroblasts were incubated only with soybean fiber

References

Abdulai A, Egger U (1992) Intraregional trade in West Africa: impacts of Ghana's cocoa exports and economic growth. Food Policy 17(2):277–286

Addai E, Gaere L (2001) Capacity-building and systems development for Sector-Wide Approaches (SWAps): the experience of the Ghana health sector. Ministry of Health, Accra

African Development Bank, OECD (2007) Country notes: Ghana. African economic outlook. OECD Publications, Paris

Ahoto Partnership for Ghana (2010) Anti-malaria project report. Unpublished. pp 1–13

Akin JS, Griffin CC, Guilkey DK et al (1985) The demand for primary health services in the third world. Rowand and Allanhel, New Jersey

Anyinam C (1991) Modern and traditional health care systems in Ghana. In: Akhtar R (ed) Health care patterns and planning in developing countries. Greenwood Press, Connecticut

Asante AD, Zwi AB (2009) Factors influencing resource allocation decisions and equity in the health system of Ghana. Public Health 123(5):371–377

Asante AD, Zwi AB, Ho MT (2006) Equity in resource allocation for health: a comparative study of the Ashanti and Northern regions of Ghana. Health Policy 78(2–3):135–148

Austrian Red Cross (2009) Health care in Ghana. Austrian Center for Country of Origin & Asylum Research and Documentation (ACCORD)

BabySoy USA (2007) http://www.babysoyusa.com. Accessed 19 Mar 2012

Balasubramaniam AD, Callister WD, Rethwisch DG (2007) Callister's materials science and engineering. Wiley India (P) Ltd, New Delhi

Barrett B (1996) Integrated local health systems in Central America. Soc Sci Med 43(1):71–82

Barrientos M (2003) Index Mundi. www.indexmundi.com/facts/ghana/rural-population. Accessed 26 Dec 2011

Belter PA, Beckel AC, Smith AK (1944) Comparison of the use of alcohol-extracted with petroleum-ether-extracted flakes in a pilot plant. Ind Eng Chem 36(9):799–803

Bhatia M, Mossialos E (2004) Health systems in developing countries. In: Hall A, Midgley J (eds) Social policy for development. Sage, London, pp 168–204

Bhatia SK, Bhatia SR (2006) Biomaterials. Encyclopedia of Chemical Processing I–V:153–160

Bhatia SK, Yetter AB (2008) Correlation of visual in vitro cytotoxicity ratings of biomaterials with quantitative in vitro cell viability measurements. Cell Biol Toxicol 24(4):315–319

Bhatia SK (2010) Biomaterials for clinical applications. Springer, New York

Birdsall N, de la Torre A, Menezes R (2008) Fair growth: economic policies for Latin America's poor and middle-income majority. Center for Global Development, Washington, DC

Bossert TJ, Beauvais JC (2002) Decentralization of health systems in Ghana, Zambia, Uganda and the Philippines: a comparative analysis of decision space. Health Policy Plan 17(1):14–31

O.A. Fatunde and S.K. Bhatia, *Medical Devices and Biomaterials for the Developing World: Case Studies in Ghana and Nicaragua*, SpringerBriefs in Public Health, DOI 10.1007/978-1-4614-4759-7, © Springer Science+Business Media New York 2012

Brooks J, Filipski M, Jonasson E et al (2011) Modelling the distributional implications of agricultural policies in developing countries: the development policy evaluation model (DEVPEM). OECD Food, Agriculture and Fisheries Working Papers, No. 50, OECD Publications, Paris

Central Intelligence Agency: The World Factbook https://www.cia.gov/library/publications/the-world-factbook. Accessed 30 Nov 2011

Cereal Food Science (2009) The structure and properties of corn. http://food--cereal.blogspot.com/2009/09/structure-and-properties-of-corn.html. Accessed 30 Nov 2011

Chin CD, Laksanasopin T, Cheung YK et al (2011) Microfluidics-based diagnostics of infectious diseases in the developing world. Nat Med 17(8):1015–1019

Coloma J, Harris E (2008) Sustainable transfer of biotechnology to developing countries: fighting poverty by bringing scientific tools to developing-country partners. Ann N Y Acad Sci 1136:358–368

Cui W, Zhou Y, Chang J (2010) Electrospun nanofibrous materials for tissue engineering and drug delivery. Sci Technol Adv Mater 11(1):1–11

Dewbre J, de Battisti AB (2008) Agricultural progress in Cameroon, Ghana and Mali: why it happened and how to sustain it. OECD Food, Agriculture and Fisheries Working Papers, No. 9, OECD Publishing

Donahue JM (1991) Planning for primary health care in Nicaragua. In: Akhtar R (ed) Health care patterns and planning in developing countries. Greenwood Press, Westport, CT

Dong J, Sun Q, Wang J (2004) Basic study of corn protein, zein, as a biomaterial in tissue engineering, surface morphology and biocompatibility. Biomaterials 25(19):4691–4697

Douglas T (2011) Biomedical engineering education in developing countries: research synthesis. Paper presented at the 33rd Annual International Conference of the IEEE EMBS, Boston, MA, 30 Aug–3 Sep 2011

Driscoll P (2009) Biopolymers in orthopedics. In: MedMarket diligence: advanced medical technologies. http://mediligence.com/blog/2009/03/27/biopolymers-orthopedics. Accessed 15 Mar 2012

Duku MH, Gu S, Hagan EB (2011) A comprehensive review of biomass resources and biofuels potential in Ghana. Renew Sustain Energy Rev 15(6):404–415

Dzogbefia VP, Arthur PL, Zakpaa HD (2007) Value addition to locally produced soybeans in Ghana: production of soy sauce using starter culture fermentation. J Sci Technol (Ghana) 27(2):22–30

Ernst RR (2006) Science, engineering, and humanity: our contribution to the future. IEEE Eng Med Biol Mag 25(3):18–19

FAOSTAT (2008) Crop production-Ghana, 2008. Food and Agriculture Organisation of the UN. http://faostat.fao.org/site/567/default.aspx. Accessed 30 Nov 2011

Food and Drug Administration. http://www.fda.gov/. Accessed 4 Oct 2011

Food and Drugs Board. http://fdbghana.gov.gh. Accessed 6 Jan 2012

Fontaine O, Garner P, Bhan MK (2007) Oral rehydration therapy: the simple solution for saving lives. BMJ 334(Suppl 1):s14

Frempong G (2009) An evaluation of the national health insurance program in Ghana. Presented at the Global Development Network (GDN) Dissemination Workshop, Pretoria, South Africa, 2–3 Jul 2009

Gaitan M, Mendez W, Sirker NE et al (1998) Growing pains: status of emergency medicine in Nicaragua. Ann Emerg Med 31(3):402–405

Galactic (2012) History of lactic acid (HLA). http://www.lactic.com/index.php/lacticacid. Accessed 14 Mar 2012

García Z (2006) Agriculture, trade negotiations, and gender. Food and Agriculture Organization (FAO) of the United Nations, Rome

Ghana Health Service (2009) 2009 GHS annual report. Ministry of Health, Accra

Giavaresi G, Fini M, Salvage J et al (2010) Bone regeneration potential of a soybean-based filler: experimental study in a rabbit cancellous bone defects. J Mater Sci Mater Med 21(2):615–626. doi:10.1007/s10856-009-3870-6

Goldman MP (2011) Cosmetic use of poly-L-lactic acid: my technique for success and minimizing complications. Dermatol Surg 37(5):688–693

Greenwood HL, Singer PA, Downey GP et al (2006) Regenerative medicine and the developing world. PLoS Med 3(9):1496–1500

Hansen J, Kaplan W, Klint K et al (2010) A stepwise approach to identify gaps in medical devices (availability matrix and survey methodology). Medical devices: managing the mismatch, Background Paper 1. World Health Organization, Geneva

Hermansson AM (1975) Functional properties of proteins for foods-flow properties. J Texture Studies 5(4):425–439

Hermansson AM (1978) Physico-chemical aspects of soy proteins structure formation. J Texture Studies 9(1–2):33–58

Heydenburg MA, Morey R (ed) (2008) Standards for medical equipment donations. International Aid-USA, pp 1–12

Horan FE (1974) Soy protein products and their production. J Am Oil Chem Soc 51(1):67A–73A

Ikada Y, Tsuji H (2000) Biodegradable polyesters for medical and ecological applications. Macromol Rapid Commun 21(3):117–132

Jamison DT, Breman JG, Measham AR et al (eds) (2006) Disease control priorities in developing countries, 2nd edn. The World Bank, Washington, DC

Kaunonen G (2010) Engineering world health: teaching people to 'fish' through biomedical engineering. IEEE Pulse 1(2):28–33

Keusch GT, Fontaine O, Bhagarva A et al (2006) Diarrheal diseases. In: Jamison DT, Breman JG, Measham AR et al (eds) Disease control priorities in developing countries, 2nd edn. IBRD/The World Bank and Oxford University Press, Washington, DC

Kim SH, Chin I-J, Yoon J-S et al (1998) Mechanical properties of biodegradable blends of poly (L-lactic acid) and starch. Korea Polymer J 6(5):422–427

Lee CH, Rha C (1977) Thickening of soy protein suspensions with calcium. J Texture Studies 7(4):441–449

Levental I, Georges PC, Janmey PA (2007) Soft biological materials and their impact on cell function. Soft Matter 3:299–306. http://pubs.rsc.org, doi:10.1039/B610522J. Accessed 26 Jan 2012

Lin BB (2011) Resilience in agriculture through crop diversification: adaptive management for environmental change. Bioscience 61(3):183–193

Linnes J (2011) Diagnostics. Lecture (unpublished). D-lab Health. Massachusetts Institute of Technology, Cambridge MA

Malhotra SV, Kumar V, East A et al (2008) Corn-based materials. In: Frontiers of engineering: reports on leading-edge engineering from the 2007 symposium. National Academy of Engineering of the National Academies. http://www.nap.edu/catalog/12027.html. Accessed 30 Jan 2012

Malkin RA (2006) Medical instrumentation in the developing world. Engineering World Health, Durham, NC

Malkin RA (2007a) Barriers for medical devices for the developing world. Expert Rev Med Devices 4(6):759–763

Malkin RA (2007b) Design of health care technologies for the developing world. Annu Rev Biomed Eng 9:567–587

Martinez AW, Phillips ST, Whitesides GM et al (2010) Diagnostics for the developing world: microfluidic paper-based analytical devices. Anal Chem 82:3–10

Mashayekh M, Mahmoodi MR, Entezari MH (2008) Effect of fortification on defatted soy flour on sensory and rheological properties of wheat bread. Int J Food Sci Technol 43:1693–1698

Meso P, Mbarika VWA, Sood SP (2009) An overview of potential factors for effective telemedicine transfer to sub-Saharan Africa. IEEE Trans Inf Technol Biomed 13(5):734–739

Mikos AG, Thorsen AJ, Czerwonka LA et al (1994) Preparation and characterization of poly(L-lactic acid) foams. Polymer 35(5):1068–1077

Ministry of Food and Agriculture, Republic of Ghana. http://mofa.gov.gh/site. Accessed 12 Nov 2011

MINSA (2009) Anexo 2: red de hospitales estatales generales. Diagnostico nacional de los servicios de atención a personas drogodependientes, Managua

Misra S, Mohanty A, Khan N (2011) Electrospinning processing parameters summary. Unpublished, pp 1–4

Moussy F (2010) Biomaterials for the developing world. J Biomed Mater Res 94(4):1001–1003

Mushtaque A, Chowdhury R, Cash RA (1996) A simple solution: teaching millions to treat diarrhea at home. University Press Ltd, Dhaka, Bangladesh

OECD (2006) OECD territorial reviews: the Mesoamerican region. OECD Publications, Paris

Osei D, d'Almeida S, George MO et al (2005) Technical efficiency of public district hospitals and health centres in Ghana: a pilot study. Cost Eff Resour Alloc 3(9). http://www.resource-alloca tion.com/content/3/1/9. Accessed 12 Dec 2011

PAHO, WHO, FDA (2001) A model regulatory program for medical devices: an international guide. Pan American Health Organization, Washington, DC

PAHO (2007) Nicaragua. In: Health in the Americas 2007, volume II-countries. Pan American Health Organization, Washington, DC

PAHO (2008) Health systems profile: Nicaragua. In: Monitoring and analyzing health systems change/reform. Pan American Health Organization, Washington, DC

Park JB (1979) Biomaterials: an introduction. Plenum, New York

Plahar WA (2006) Overview of the soybean industry in Ghana. Presented at the Workshop on soybean protein for human nutrition and health, Accra, Ghana, 28 Sept 2006

Poon Carmen CY, Zhang Y-T (2008) Perspectives on high technologies for low-cost healthcare: the Chinese scenario. IEEE Eng Med Biol Mag 27(5):42–47

Program in Infectious Disease and Social Change, Harvard Medical School (1999) Chapter 1: the global impact of drug-resistant tuberculosis. Harvard Medical School and the Open Society Institute, Boston

Qi G, Venkateshan K, Mo X et al (2011) Physicochemical properties of soy protein: effects of subunit composition. J Agric Food Chem 59(18):9958–9964

Ratner BD, Bryant SJ (2004) Biomaterials: where we have been and where we are going. Annu Rev Biomed Eng 6:41–75

Rea L, Bhatia SK (2012) Mechanical characterization of corn-derived Poly-L-lactic acid Presented at the 38th annual northeast bioengineering conference, Temple University, Philadelphia, PA, 16–18 Mar 2012, pp 175–176

Salisu A, Prinz V, Yoshimura D (ed) (2009) Health Care in Ghana. ACCORD/Austrian Red Cross, Vienna

Sandhu SS, Singh N (2007) Some properties of corn starches II: physicochemical, gelatinization, retrogradation, pasting, and gel textural properties. Food Chem 101:1499–1507

San Jose State University (1997) Exploring materials engineering: biomaterials. www.engr.sjsu. edu/WofMatE/Biomaterials.htm. Accessed 16 Oct 2011

Scott JAG, Hall AJ, Muyodi C et al (2000) Aetiology, outcome, and risk factors for mortality among adults with acute pneumonia in Kenya. Lancet 355(9211):1225–1230

Sinha SR, Barry M (2011) Health technologies and innovation in the global health arena. N Engl J Med 365(9):779–782

SODI (2011) German-Nicaraguan hospital in Managua. http://www.sodi.de/en/projects/nicaragua/ german_nicaraguan_hospital_in_managua. Accessed 29 Mar 2012

Soliman E, Yang SC, Dombi GW et al (2012) Electrically conductive, biocompatible composite containing carbon nanobrushes for applications in neuroregeneration. Presented at the 38th annual northeast bioengineering conference, Philadelphia, PA, 16–18 Mar 2012

Texas Tech University Department of Plant and Soil Science (2012) Seed to seedling: soybeans. http://www.itc.ttu.edu/pss1321/web%20topics/seedtoseedlingnew.htm. Accessed 4 Jan 2012

Turshen M (1989) Disease eradication. In: The politics of public health. Rutgers University Press, New Brunswick

Twumasi PA (1981) Colonialism and international health: a study in social change in Ghana. Soc Sci Med 15B:147–151

Uchimura H (ed) (2009) Making health services more accessible in developing countries: finance and health resources for functioning health systems. IDE-JETRO, London

UNICEF (2005) Primary health centres and first referral level hospitals planning guide: equipment and renewable resources. UNICEF

Vega-Lugo AC, Lim LT (2009) Controlled release of allyl isothiocyanate using soy protein and poly(lactic acid) electrospun fibers. Food Res Int 42(8):933–940

Washington Life Science (2012) Definition of medical technology. http://www.washingtonlifescience.com/industry/definition_medtech.htm. Accessed 28 Aug 2011

Wheeler DA, Arathoon EG, Pitts M et al (2001) Availability of HIV care in Central America. JAMA 286(7):853–860

World Health Rankings (2012) http://www.worldlifeexpectancy.com. Accessed 11 Jan 2012

WHO (2003) Medical device regulations: global overview and guiding principles. World Health Organization, Geneva

WHO (2006) The stop TB strategy: building on and enhancing DOTS to meet the TB-related millennium development goals. World Health Organization, Geneva

WHO, John Snow Inc, PATH et al (2008) Interagency list of essential medical devices for reproductive health. World Health Organization, Geneva

WHO (2010a) Appendix A: criteria for medical equipment inventory inclusion. In: Introduction to medical equipment inventory management. WHO Medical device technical series. World Health Organization, Geneva

WHO (2010b) Barriers to innovation in the field of medical devices. Background paper 6. In: Medical devices: managing the mismatch. World Health Organization, Geneva

WHO (2010c) Baseline country survey on medical devices. World Health Organization, Geneva

WHO (2010d) Future public health needs: commonalities and differences between high- and low-resource settings. Background paper 8. In: Medical devices: managing the mismatch. World Health Organization: Geneva

WHO (2011a) Medical devices by health care facility (in progress). World Health Organization, Geneva

WHO (2011b) Needs assessment for medical devices. WHO Medical device technical series. World Health Organization, Geneva

WHO (2011c) World health statistics 2011. World Health Organization, Geneva

WHO (2012) Global health observatory. www.who.int/gho. Accessed 30 March 2012

Yager P, Edwards T, Fu E et al (2006) Microfluidic diagnostic technologies for global public health. Nature 442(7101):412–418

Zahedi E (2011) The dilemma of BME research projects in developing countries: a case study. Presented at the 33rd annual international conference of the IEEE EMBS, Boston, MA, 30 Aug–3 Sept 2011

Index

O.A. Fatunde and S.K. Bhatia, *Medical Devices and Biomaterials for the Developing World:*   111
Case Studies in Ghana and Nicaragua, SpringerBriefs in Public Health,
DOI 10.1007/978-1-4614-4759-7, © Springer Science+Business Media New York 2012